1

An Introduction to Crystallography

Crystallography is the study of the arrangement of atoms in a crystalline solid. This chapter discusses briefly all the important types of crystallographic methods such as quantum crystallography, x-ray crystallography, electron crystallography, nuclear magnetic resonance crystallography, etc.

Crystallography is the science that examines crystals, which can be found everywhere in nature—from salt to snowflakes to gemstones. Crystallographers use the properties and inner structures of crystals to determine the arrangement of atoms and generate knowledge that is used by chemists, physicists, biologists, and others. Within the past century, crystallography has been a primary force in driving major advances in the detailed understanding of materials, synthetic chemistry, the understanding of basic principles of biological processes, genetics, and has contributed to major advances in the development of drugs for numerous diseases. As a science, crystallography has produced 28 Nobel Prizes, more than any other scientific field.

Crystallographers use X-ray, neutron, and electron diffraction techniques to identify and characterize solid materials. They commonly bring in information from other analytical techniques, including X-ray fluorescence, spectroscopic techniques, microscopic imaging, and computer modeling and visualization to construct detailed models of the atomic arrangements in solids. This provides valuable information on a material's chemical makeup, polymorphic form, defects or disorder, and electronic properties. It also sheds light on how solids perform under temperature, pressure, and stress conditions.

Crystal-growing specialists use a variety of techniques to produce crystalline forms of compounds for use in research or manufacturing. They may be experts in working with hard-to-crystallize materials, or they may grow crystals to exacting specifications for use in computer chips, solar cells, optical components, or pharmaceutical products.

Single-crystal X-ray crystallography is widely considered to be the gold standard for establishing the structures of crystalline solids. This method is used to establish patent claims, establish structure-property relationships for new compounds, and many other applications. However, powder crystallography instrumentation and data analysis software have emerged over the past 30 years as powerful methods for investigating the structures of materials that cannot be studied with single-crystal methods. Powder methods are used in a wide variety of investigations, including forensic analyses, identifying components of mixtures, and identifying properties of polymers and other poorly crystallized materials.

The pharmaceutical and biochemical fields rely extensively on crystallographic studies. Proteins and other biological materials (including viruses) may be crystallized to aid in studying their structures and composition. Many important pharmaceuticals are administered in crystalline form, and detailed descriptions of their crystal structures provide evidence to verify claims in patents.

Instrument manufacturers hire crystallographers for customer sales and support functions, including instrument repair and helping customers with special projects. Staff crystallographers at the national laboratories develop and maintain leading-edge research instruments and software capabilities. They also assist visiting users in setting up and running experiments using specialized techniques, including synchrotron X-ray diffraction and neutron diffraction. Universities employ staff members to maintain and operate their research laboratories and to train students to use the instruments.

Some crystallographers develop instrumentation and software for collecting, analyzing, and visualizing data and for translating this data into crystal structure models. Some crystallographers maintain and develop archival databases at industrial and academic institutions, as well as some nonprofits and government laboratories.

Service laboratories hire diffraction technicians to prepare and catalog samples, run the data collections, and prepare routine reports on the results. Technicians may also be called on to perform routine instrument maintenance and simple repairs.

Forensics laboratories use crystallography to investigate cases involving product adulteration or counterfeiting. They may identify minerals, metals, or other materials found at crime scenes. They may also identify corrosion products and other residues found at the site of an industrial accident to help verify the events leading up to the accident.

Theory

An image of a small object is made using a lens to focus the beam, similar to a lens in a microscope. However, the wavelength of visible light (about 4000 to 7000 ångström) is three orders of magnitude longer than the length of typical atomic bonds and atoms themselves (about 1 to 2 Å). Therefore, obtaining information about the spatial arrangement of atoms requires the use of radiation with shorter wavelengths, such as X-ray or neutron beams. Employing shorter wavelengths implied abandoning microscopy and true imaging, however, because there exists no material from which a lens capable of focusing this type of radiation can be created. Scientists have had some success focusing X-rays with microscopic Fresnel zone plates made from gold, and by critical-angle reflection inside long tapered capillaries. Diffracted X-ray or neutron beams cannot be focused to produce images, so the sample structure must be reconstructed from the diffraction pattern. Sharp features in the diffraction pattern arise from periodic, repeating structure in the sample, which are often very strong due to coherent reflection of many photons from many regularly spaced instances of similar structure, while non-periodic components of the structure result in diffuse (and usually weak) diffraction features - areas with a higher density and repetition of atom order tend to reflect more light toward one point in space when compared to those areas with fewer atoms and less repetition.

Because of their highly ordered and repetitive structure, crystals give diffraction patterns of sharp Bragg reflection spots, and are ideal for analyzing the structure of solids.

Notation

- Coordinates in *square brackets* such as [100] denote a direction vector (in real space).

- Coordinates in *angle brackets* or *chevrons* such as <100> denote a *family* of directions which are related by symmetry operations. In the cubic crystal system for example, <100> would mean [100], [010], [001] or the negative of any of those directions.

- Miller indices in *parentheses* such as (100) denote a plane of the crystal structure, and regular repetitions of that plane with a particular spacing. In the cubic system, the normal to the (hkl) plane is the direction [hkl], but in lower-symmetry cases, the normal to (hkl) is not parallel to [hkl].

- Indices in *curly brackets* or *braces* such as {100} denote a family of planes and their normals which are equivalent in cubic materials due to symmetry operations, much the way angle brackets denote a family of directions. In non-cubic materials, <hkl> is not necessarily perpendicular to {hkl}.

Techniques

Some materials that have been analyzed crystallographically, such as proteins, do not occur naturally as crystals. Typically, such molecules are placed in solution and allowed to slowly crystallize through vapor diffusion. A drop of solution containing the molecule, buffer, and precipitants is sealed in a container with a reservoir containing a hygroscopic solution. Water in the drop diffuses to the reservoir, slowly increasing the concentration and allowing a crystal to form. If the concentration were to rise more quickly, the molecule would simply precipitate out of solution, resulting in disorderly granules rather than an orderly and hence usable crystal.

Once a crystal is obtained, data can be collected using a beam of radiation. Although many universities that engage in crystallographic research have their own X-ray producing equipment, synchrotrons are often used as X-ray sources, because of the purer and more complete patterns such sources can generate. Synchrotron sources also have a much higher intensity of X-ray beams, so data collection takes a fraction of the time normally necessary at weaker sources. Complementary neutron crystallography techniques are used to identify the positions of hydrogen atoms, since X-rays only interact very weakly with light elements such as hydrogen.

Producing an image from a diffraction pattern requires sophisticated mathematics and often an iterative process of modelling and refinement. In this process, the mathematically predicted diffraction patterns of an hypothesized or "model" structure are compared to the actual pattern generated by the crystalline sample. Ideally, researchers make several initial guesses, which through refinement all converge on the same answer. Models are refined until their predicted patterns match to as great a degree as can be achieved without radical revision of the model. This is a painstaking process, made much easier today by computers.

The mathematical methods for the analysis of diffraction data only apply to *patterns,* which in turn result only when waves diffract from orderly arrays. Hence crystallography applies for the most part only to crystals, or to molecules which can be coaxed to crystallize for the sake of measurement. In spite of this, a certain amount of molecular information can be deduced from patterns that are generated by fibers and powders, which while not as perfect as a solid crystal, may exhibit a degree of order. This level of order can be sufficient to deduce the structure of simple molecules, or to determine the coarse features of more complicated molecules. For example, the double-helical structure of DNA was deduced from an X-ray diffraction pattern that had been generated by a fibrous sample.

In Materials Science

Crystallography is used by materials scientists to characterize different materials. In single crystals, the effects of the crystalline arrangement of atoms is often easy to see macroscopically, because the natural shapes of crystals reflect the atomic structure. In addition, physical properties are often controlled by crystalline defects. The understanding of crystal structures is an important prerequisite for understanding crystallographic defects. Mostly, materials do not occur as a single crystal, but in poly-crystalline form (i.e., as an aggregate of small crystals with different orientations). Because of this, the powder diffraction method, which takes diffraction patterns of polycrystalline samples with a large number of crystals, plays an important role in structural determination.

Other physical properties are also linked to crystallography. For example, the minerals in clay form small, flat, platelike structures. Clay can be easily deformed because the platelike particles can slip along each other in the plane of the plates, yet remain strongly connected in the direction perpendicular to the plates. Such mechanisms can be studied by crystallographic texture measurements.

In another example, iron transforms from a body-centered cubic (bcc) structure to a face-centered cubic (fcc) structure called austenite when it is heated. The fcc structure is a close-packed structure unlike the bcc structure; thus the volume of the iron decreases when this transformation occurs.

Crystallography is useful in phase identification. When manufacturing or using a material, it is generally desirable to know what compounds and what phases are present in the material, as their composition, structure and proportions will influence the material's properties. Each phase has a characteristic arrangement of atoms. X-ray or neutron diffraction can be used to identify which patterns are present in the material, and thus which compounds are present. Crystallography covers the enumeration of the symmetry patterns which can be formed by atoms in a crystal and for this reason is related to group theory and geometry.

Biology

X-ray crystallography is the primary method for determining the molecular conformations of biological macromolecules, particularly protein and nucleic acids such as DNA and RNA. In fact, the double-helical structure of DNA was deduced from crystallographic data. The first crystal structure of a macromolecule was solved in 1958, a three-dimensional model of the myoglobin molecule obtained by X-ray analysis. The Protein Data Bank (PDB) is a freely accessible repository for the structures of proteins and other biological macromolecules. Computer programs such as RasMol or Pymol can be used to visualize biological molecular structures.Neutron crystallography is often used to help refine structures obtained by X-ray methods or to solve a specific bond; the methods are often viewed as complementary, as X-rays are sensitive to electron positions and scatter most strongly off heavy atoms, while neutrons are sensitive to nucleus positions and scatter strongly even off many light isotopes, including hydrogen and deuterium.Electron crystallography has been used to determine some protein structures, most notably membrane proteins and viral capsids.

Quantum Crystallography

Quantum crystallography (QCr) is a branch of crystallography aimed at obtaining the complete quantum mechanics of a crystal given its X-ray scattering data. The fundamental value of obtaining an electron density matrix that is N-representable is that it ensures consistency with an underlying properly antisymmetrized wavefunction, a requirement of quantum mechanical validity. However, X-ray crystallography has progressed in an impressive way for decades based only upon the electron density obtained from the X-ray scattering data without the imposition of the mathematical structure of quantum mechanics.

The term quantum crystallography (QCr) refers to the combination of structural, mainly crystallographic, information with quantum-mechanical theory.

Quantum chemistry methods and crystal structure determination are highly developed research tools, indispensable in today's organic, inorganic and physical chemistry. These tools are usually employed separately. Diffraction and scattering experiments provide structures at the atomic scale, while the techniques of quantum chemistry provide wavefunctions and properties derived from them. In this contribution we review different efforts towards combining tools of these two fields into integrated quantum crystallographic strategies.

The term quantum crystallography was first introduced by Massa, Huang and Karle in 1995 for methods that exploit "crystallographic information to enhance quantum mechanical calculations and the information derived from them". The basic idea is to compensate shortcomings and limitations of quantum mechanical models, e.g. incomplete consideration of electron correlation, with experimental data that are not suffering from the same limitations (Fig. a). In 1999 the same authors also suggested the converse possibility: "quantum mechanics can greatly enhance the information available from a crystallographic experiment" (Fig. b).

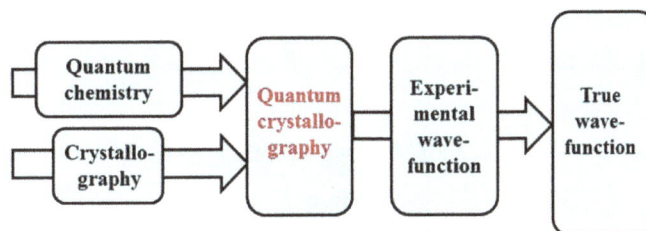

Fig. a

Quantum crystallography, first aspect: crystallographic data are integrated into quantum chemical calculations to enhance the information content of the wavefunction. The resulting, so-called "experimental wavefunction" represents an improved approximation to the true wavefunction.

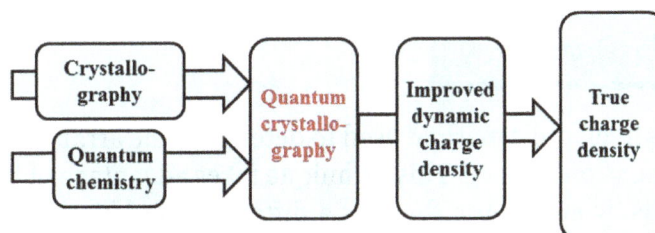

Fig. b

Quantum crystallography, second aspect: quantum chemical calculations are integrated into crystal structure determination to improve the dynamic charge density, *i.e.* the thermally smeared electron and nuclear densities.

Mutually subsidiary combinations of quantum chemistry and X-ray structure determination suggest themselves. Quantum chemical models are usually based on some approximations of the true wavefunction. X-ray structure determination aims at the true charge density, which is related to the square of the true wavefunction, but is affected to a smaller or larger extent by vibrational motion and experimental errors.

Here the topic of quantum crystallography is presented in two parts. The first part summarizes the development of increasingly sophisticated methods to combine information from quantum chemical calculations with diffraction and other experimental data (Fig. c for a summarizing scheme). The second part describes ways of improving structural models obtained from diffraction experiments by combining them in a self-consistent way with information from quantum chemical calculations.

QUANTUM CHEMISTRY

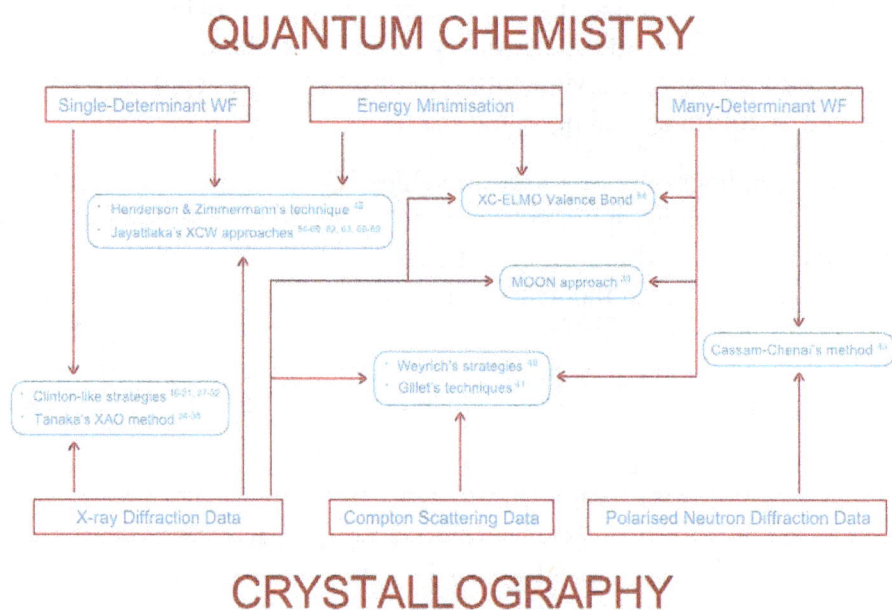

Fig. c

Scheme summarizing the features (framed in red) of the main methods (framed in light blue) according to the first definition of quantum crystallography. The lower the position of the method is in the scheme, the lower is the quantum chemistry contribution in it.

X-ray Crystallography

X-ray Crystallography is a scientific method used to determine the arrangement of atoms of a crystalline solid in three dimensional space. This technique takes advantage of the interatomic spacing of most crystalline solids by employing them as a diffraction gradient for x-ray light, which has wavelengths on the order of 1 angstrom (10^{-8} cm).

Introduction

In 1895, Wilhelm Rontgen discovered x- rays. The nature of x- rays, whether they were particles or electromagnetic radiation, was a topic of debate until 1912. If the wave idea was correct, researchers knew that the wavelength of this light would need to be on the order of 1 Angstrom (A) (10^{-8} cm). Diffraction and measurement of such small wavelengths would require a gradient with spacing on the same order of magnitude as the light.

In 1912, Max von Laue, at the University of Munich in Germany, postulated that atoms in a crystal lattice had a regular, periodic structure with interatomic distances on the order of 1 A. Without having any evidence to support his claim on the periodic arrangements of atoms in a lattice, he further postulated that the crystalline structure can be used to diffract x-rays, much like a gradient in an infrared spectrometer can diffract infrared light. His postulate was based on the following assumptions: the atomic lattice of a crystal is periodic, x- rays are electromagnetic radiation, and the interatomic distance of a crystal are on the same order of magnitude as x- ray light. Laue's predictions were confirmed when two researchers: Friedrich and Knipping, successfully photographed the diffraction pattern associated with the x-ray radiation of crystalline $CuSO_4 \cdot 5H_2O$. The science of x-ray crystallography was born.

The arrangement of the atoms needs to be in an ordered, periodic structure in order for them to diffract the x-ray beams. A series of mathematical calculations is then used to produce a diffraction pattern that is characteristic to the particular arrangement of atoms in that crystal. X-ray crystallography remains to this day the primary tool used by researchers in characterizing the structure and bonding of organometallic compounds.

Diffraction

Diffraction is a phenomena that occurs when light encounters an obstacle. The waves of light can either bend around the obstacle, or in the case of a slit, can travel through the slits. The resulting diffraction pattern will show areas of constructive interference, where two waves interact in phase, and destructive interference, where two waves interact out of phase. Calculation of the phase difference can be explained by examining Figure below.

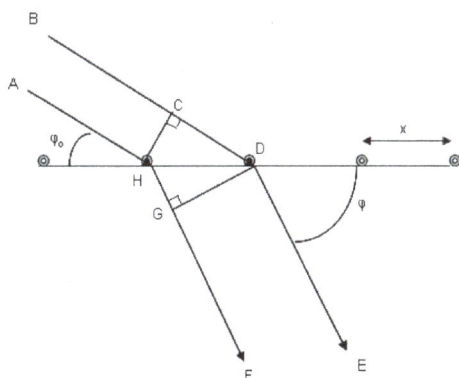

Scattering of light by a diffraction gratting

In the figure below, two parallel waves, BD and AH are striking a gradient at an angle θ_0. The incident wave BD travels farther than AH by a distance of CD before reaching the gradient. The scattered wave (depicted below the gradient) HF, travels father than the scattered wave DE by a distance of HG. So the total path difference between path AHGF and BCDE is CD - HG. To observe a wave of high intensity (one created through constructive interference), the difference CD - HG must equal to an integer number of wavelengths to be observed at the angle psi, CD–HG=nλ, where λ is the wavelength of the light. Applying some basic trigonometric properties, the following two equations can be shown about the lines:

CD=xcos(θo)

and

HG=xcos(θ)

where x is the distance between the points where the diffraction repeats. Combining the two equations,

x(cosθo–cosθ)=nλ

Diffraction Theory

The main goal of X-ray crystallography is to determine the density of electrons $f(r)$ throughout the crystal, where r represents the three-dimensional position vector within the crystal. To do this, X-ray scattering is used to collect data about its Fourier transform $F(q)$, which is inverted mathematically to obtain the density defined in real space, using the formula

$$f(\mathbf{r}) = \frac{1}{(2\pi)^3} \int F(\mathbf{q}) e^{i\mathbf{q}\cdot\mathbf{r}} d\mathbf{q}$$

where the integral is taken over all values of q. The three-dimensional real vector q represents a point in reciprocal space, that is, to a particular oscillation in the electron density as one moves in the direction in which q points. The length of q corresponds to 2π divided by the wavelength of the oscillation. The corresponding formula for a Fourier transform will be used below

$$F(\mathbf{q}) = \int f(\mathbf{r}) e^{-i\mathbf{q}\cdot\mathbf{r}} d\mathbf{r}$$

where the integral is summed over all possible values of the position vector r within the crystal.

The Fourier transform $F(q)$ is generally a complex number, and therefore has a magnitude $|F(q)|$ and a phase $\varphi(q)$ related by the equation

$$F(\mathbf{q}) = |F(\mathbf{q})| e^{i\phi(\mathbf{q})}$$

The intensities of the reflections observed in X-ray diffraction give us the magnitudes $|F(q)|$ but not the phases $\varphi(q)$. To obtain the phases, full sets of reflections are collected with known alterations to the scattering, either by modulating the wavelength past a certain absorption edge or by adding strongly scattering (i.e., electron-dense) metal atoms such as mercury. Combining the magnitudes and phases yields the full Fourier transform $F(q)$, which may be inverted to obtain the electron density $f(r)$.

Crystals are often idealized as being *perfectly* periodic. In that ideal case, the atoms are positioned on a perfect lattice, the electron density is perfectly periodic, and the Fourier transform $F(q)$ is zero except when q belongs to the reciprocal lattice (the so-called *Bragg peaks*). In reality, however, crystals are not perfectly periodic; atoms vibrate about their mean position, and there may be disorder of various types, such as mosaicity, dislocations, various point defects, and heterogeneity in the conformation of crystallized molecules. Therefore, the Bragg peaks have a finite width and there may be significant *diffuse scattering*, a continuum of scattered X-rays that fall between the Bragg peaks.

Intuitive Understanding by Bragg's Law

An intuitive understanding of X-ray diffraction can be obtained from the Bragg model of diffraction. In this model, a given reflection is associated with a set of evenly spaced sheets running through the crystal, usually passing through the centers of the atoms of the crystal lattice. The orientation of a particular set of sheets is identified by its three Miller indices (h, k, l), and let their spacing be noted by d. William Lawrence Bragg proposed a model in which the incoming X-rays are scattered specularly (mirror-like) from each plane; from that assumption, X-rays scattered from adjacent planes will combine constructively (constructive interference) when the angle θ between the plane and the X-ray results in a path-length difference that is an integer multiple n of the X-ray wavelength λ.

$$2d \sin \theta = n\lambda$$

A reflection is said to be *indexed* when its Miller indices (or, more correctly, its reciprocal lattice vector components) have been identified from the known wavelength and the scattering angle 2θ. Such indexing gives the unit-cell parameters, the lengths and angles of the unit-cell, as well as its space group. Since Bragg's law does not interpret the relative intensities of the reflections, however, it is generally inadequate to solve for the arrangement of atoms within the unit-cell; for that, a Fourier transform method must be carried out.

Scattering as a Fourier transform

The incoming X-ray beam has a polarization and should be represented as a vector wave; however, for simplicity, let it be represented here as a scalar wave. We also ignore the complication of the time dependence of the wave and just concentrate on the wave's spatial dependence. Plane waves can be represented by a wave vector k_{in}, and so the strength of the incoming wave at time $t=0$ is given by

$$A e^{i\mathbf{k}_{in} \cdot \mathbf{r}}$$

At position r within the sample, let there be a density of scatterers $f(r)$; these scatterers should produce a scattered spherical wave of amplitude proportional to the local amplitude of the incoming wave times the number of scatterers in a small volume dV about r.

$$\text{amplitude of scattered wave} = A e^{i\mathbf{k} \cdot \mathbf{r}} S f(\mathbf{r}) dV$$

where S is the proportionality constant.

Let's consider the fraction of scattered waves that leave with an outgoing wave-vector of k_{out} and strike the screen at r_{screen}. Since no energy is lost (elastic, not inelastic scattering), the wavelengths are the same as are the magnitudes of the wave-vectors $|k_{in}| = |k_{out}|$. From the time that the photon is scattered at r until it is absorbed at r_{screen}, the photon undergoes a change in phase

$$e^{ik_{out} \cdot (r_{screen} - r)}$$

The net radiation arriving at r_{screen} is the sum of all the scattered waves throughout the crystal

$$AS \int dr f(\mathbf{r}) e^{ik_{in} \cdot r} e^{ik_{out} \cdot (r_{screen} - r)} = ASe^{ik_{out} \cdot r_{screen}} \int dr f(\mathbf{r}) e^{i(k_{in} - k_{out}) \cdot r}$$

which may be written as a Fourier transform

$$ASe^{ik_{out} \cdot r_{screen}} \int dr f(\mathbf{r}) e^{-iq \cdot r} = ASe^{ik_{out} \cdot r_{screen}} F(\mathbf{q})$$

where $q = k_{out} - k_{in}$. The measured intensity of the reflection will be square of this amplitude

$$A^2 S^2 |F(\mathbf{q})|^2$$

Friedel and Bijvoet Mates

For every reflection corresponding to a point q in the reciprocal space, there is another reflection of the same intensity at the opposite point -q. This opposite reflection is known as the *Friedel mate* of the original reflection. This symmetry results from the mathematical fact that the density of electrons $f(r)$ at a position r is always a real number. As noted above, $f(r)$ is the inverse transform of its Fourier transform $F(q)$; however, such an inverse transform is a complex number in general. To ensure that $f(r)$ is real, the Fourier transform $F(q)$ must be such that the Friedel mates $F(-q)$ and $F(q)$ are complex conjugates of one another. Thus, $F(-q)$ has the same magnitude as $F(q)$ but they have the opposite phase, i.e., $\varphi(q) = -\varphi(q)$

$$F(-\mathbf{q}) = |F(-\mathbf{q})| e^{i\phi(-q)} = F^*(\mathbf{q}) = |F(\mathbf{q})| e^{-i\phi(q)}$$

The equality of their magnitudes ensures that the Friedel mates have the same intensity $|F|^2$. This symmetry allows one to measure the full Fourier transform from only half the reciprocal space, e.g., by rotating the crystal slightly more than 180° instead of a full 360° revolution. In crystals with significant symmetry, even more reflections may have the same intensity (Bijvoet mates); in such cases, even less of the reciprocal space may need to be measured. In favorable cases of high symmetry, sometimes only 90° or even only 45° of data are required to completely explore the reciprocal space.

The Friedel-mate constraint can be derived from the definition of the inverse Fourier transform

$$f(\mathbf{r}) = \int \frac{dq}{(2\pi)^3} F(\mathbf{q}) e^{iq \cdot r} = \int \frac{dq}{(2\pi)^3} |F(\mathbf{q})| e^{i\phi(q)} e^{iq \cdot r}$$

Since Euler's formula states that $e^{ix} = \cos(x) + i \sin(x)$, the inverse Fourier transform can be separated into a sum of a purely real part and a purely imaginary part

$$f(\mathbf{r}) = \int \frac{d\mathbf{q}}{(2\pi)^3} |F(\mathbf{q})| e^{i(\phi + \mathbf{q} \cdot \mathbf{r})}$$

$$= \int \frac{d\mathbf{q}}{(2\pi)^3} |F(\mathbf{q})| \cos(\phi + \mathbf{q} \cdot \mathbf{r}) + i \int \frac{d\mathbf{q}}{(2\pi)^3} |F(\mathbf{q})| \sin(\phi + \mathbf{q} \cdot \mathbf{r}) = I_{cos} + i I_{sin}$$

The function $f(\mathbf{r})$ is real if and only if the second integral I_{sin} is zero for all values of r. In turn, this is true if and only if the above constraint is satisfied

$$I_{sin} = \int \frac{d\mathbf{q}}{(2\pi)^3} |F(\mathbf{q})| \sin(\phi + \mathbf{q} \cdot \mathbf{r}) = \int \frac{d\mathbf{q}}{(2\pi)^3} |F(-\mathbf{q})| \sin(-\phi - \mathbf{q} \cdot \mathbf{r}) = -I_{sin}$$

since $I_{sin} = -I_{sin}$ implies that $I_{sin} = 0$.

Ewald's Sphere

Each X-ray diffraction image represents only a slice, a spherical slice of reciprocal space, as may be seen by the Ewald sphere construction. Both k_{out} and k_{in} have the same length, due to the elastic scattering, since the wavelength has not changed. Therefore, they may be represented as two radial vectors in a sphere in reciprocal space, which shows the values of q that are sampled in a given diffraction image. Since there is a slight spread in the incoming wavelengths of the incoming X-ray beam, the values of $|F(\mathbf{q})|$ can be measured only for q vectors located between the two spheres corresponding to those radii. Therefore, to obtain a full set of Fourier transform data, it is necessary to rotate the crystal through slightly more than 180°, or sometimes less if sufficient symmetry is present. A full 360° rotation is not needed because of a symmetry intrinsic to the Fourier transforms of real functions (such as the electron density), but "slightly more" than 180° is needed to cover all of reciprocal space within a given resolution because of the curvature of the Ewald sphere. In practice, the crystal is rocked by a small amount (0.25-1°) to incorporate reflections near the boundaries of the spherical Ewald's shells.

Patterson Function

A well-known result of Fourier transforms is the autocorrelation theorem, which states that the autocorrelation c(r) of a function $f(\mathbf{r})$

$$c(\mathbf{r}) = \int d\mathbf{x} f(\mathbf{x}) f(\mathbf{x} + \mathbf{r}) = \int \frac{d\mathbf{q}}{(2\pi)^3} C(\mathbf{q}) e^{i\mathbf{q} \cdot \mathbf{r}}$$

has a Fourier transform C(q) that is the squared magnitude of F(q)

$$C(\mathbf{q}) = |F(\mathbf{q})|^2$$

Therefore, the autocorrelation function c(r) of the electron density (also known as the *Patterson function*) can be computed directly from the reflection intensities, without computing the phases.

In principle, this could be used to determine the crystal structure directly; however, it is difficult to realize in practice. The autocorrelation function corresponds to the distribution of vectors between atoms in the crystal; thus, a crystal of N atoms in its unit cell may have $N(N-1)$ peaks in its Patterson function. Given the inevitable errors in measuring the intensities, and the mathematical difficulties of reconstructing atomic positions from the interatomic vectors, this technique is rarely used to solve structures, except for the simplest crystals.

Advantages of a Crystal

In principle, an atomic structure could be determined from applying X-ray scattering to non-crystalline samples, even to a single molecule. However, crystals offer a much stronger signal due to their periodicity. A crystalline sample is by definition periodic; a crystal is composed of many unit cells repeated indefinitely in three independent directions. Such periodic systems have a Fourier transform that is concentrated at periodically repeating points in reciprocal space known as *Bragg peaks*; the Bragg peaks correspond to the reflection spots observed in the diffraction image. Since the amplitude at these reflections grows linearly with the number N of scatterers, the observed *intensity* of these spots should grow quadratically, like N^2. In other words, using a crystal concentrates the weak scattering of the individual unit cells into a much more powerful, coherent reflection that can be observed above the noise. This is an example of constructive interference.

In a liquid, powder or amorphous sample, molecules within that sample are in random orientations. Such samples have a continuous Fourier spectrum that uniformly spreads its amplitude thereby reducing the measured signal intensity, as is observed in SAXS. More importantly, the orientational information is lost. Although theoretically possible, it is experimentally difficult to obtain atomic-resolution structures of complicated, asymmetric molecules from such rotationally averaged data. An intermediate case is fiber diffraction in which the subunits are arranged periodically in at least one dimension.

Crystallization

In order to run an x-ray diffraction experiment, one must first obtain a crystal. In organometallic chemistry, a reaction might work but when no crystals form, it is impossible to characterize the products. Crystals are grown by slowly cooling a supersaturated solution. Such a solution can be made by heating a solution to decrease the amount of solvent present and to increase the solubility of the desired compound in the solvent. Once made, the solution must be cooled gradually. Rapid temperature change will cause the compound to crash out of solution, trapping solvent and impurities within the newly formed matrix. Cooling continues as a seed crystal forms. This crystal is a point where solute can deposit out of the solution and into the solid phase. Solutions are generally placed into a freezer (-78 °C) in order to ensure all of the compound has crystallized. One way to ensure gradual cooling in a -78 °C freezer is to place the container housing the compound into a beaker of ethanol. The ethanol will act as a temperature buffer, ensuring a slow decrease in the temperature gradient between the flask and the freezer. Once crystals are grown, it is imperative that they remain cold as any addition of energy will cause a disruption of the crystal lattice, which will yield bad diffraction data. The result of an organometallic chromium compound crystallization can be seen below.

Organometallic chromium crystals in a Schlenk under nitrogen

Mounting the Crystal

Due to the air-sensitivity of most organometallic compounds, crystals must be transported in a highly viscous organic compound called paratone oil. Crystals are abstracted from their respective Schlenks by dabbing the end of a spatula with the paratone oil and then sticking the crystal onto the oil. Although there might be some exposure of the compounds to air and water, crystals can withstand more exposure than solution (of the preserved protein) before degrading. On top of serving to protect the crystal, the paratone oil also serves as the glue to bind the crystal to the needle.

Rotating Crystal Method

To describe the periodic, three dimensional nature of crystals, the Laue equations are employed:

$$a(\cos\theta_0 - \cos\theta) = h\lambda$$

$$b(\cos\theta_0 - \cos\theta) = k\lambda$$

$$c(\cos\theta_0 - \cos\theta) = l\lambda$$

where aa, bb, and cc are the three axes of the unit cell, θo, o,?o are the angles of incident radiation, and ?, ?, ? are the angles of the diffracted radiation. A diffraction signal (constructive interference) will arise when h, k, and l are integer values. The rotating crystal method employs these equations. X-ray radiation is shown onto a crystal as it rotates around one of its unit cell axis. The beam strikes the crystal at a 90 degree angle. Using equation 1 above, we see that if θo is 90 degrees, then $\cos\theta$o=0. For the equation to hold true, we can set h=0, granted that \theta= 90. The above three equations will be satisfied at various points as the crystal rotates. This gives rise to a diffraction pattern (shown in the image below as multiple h values). The cylindrical film is then unwrapped and developed. The following equation can be used to determine the length axis around which the crystal was rotated:

$$a = \frac{ch\lambda}{\sin\tan^{-1}(y/r}$$

where a is the length of the axis, y is the distance from h=0 to the h of interest, r is the radius of the firm, and ? is the wavelength of the x-ray radiation used. The first length can be determined with ease, but the other two require far more work, including remounting the crystal so that it rotates around that particular axis.

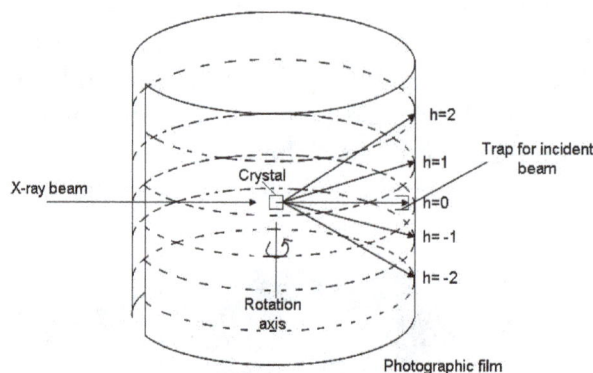

X-ray Crystallography of Proteins

The crystals that form are frozen in liquid nitrogen and taken to the synchrotron which is a highly powered tunable x-ray source. They are mounted on a goniometer and hit with a beam of x-rays. Data is collected as the crystal is rotated through a series of angles. The angle depends on the symmetry of the crystal.

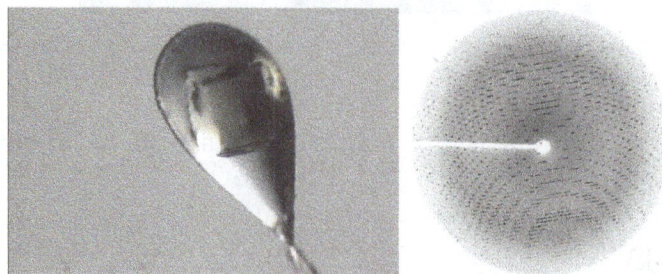

Top Left) This is a picture of a protein crystal mounted on a loop with respect to the UC Davis Structural Biology Lab; Bottom Right) This is a diffraction pattern created from the APS Kinase D63N Mutant of the above crystal with respect to the UC Davis Structural Biology Lab

Proteins are among the many biological molecules that are used for x-ray Crystallography studies. They are involved in many pathways in biology, often catalyzing reactions by increasing the reaction rate. Most scientists use x-ray Crystallography to solve the structures of protein and to determine functions of residues, interactions with substrates, and interactions with other proteins or nucleic acids. Proteins can be co - crystallized with these substrates, or they may be soaked into the crystal after crystallization.

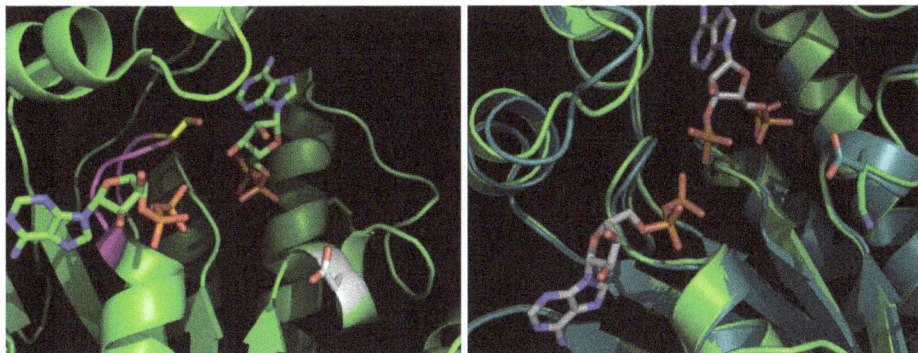

Top Left) This is the structure of APS Kinase co - crystallized with ligands ADP and APS created via pymol by an undergrad working in the Structural Biology lab at UC Davis; Bottom right).

This is the mutant overlay of APS kinase. The teal is the wild - type and the lime green is the mutant. D63 (from the wild-type) is mutated to asparagine. Images created by pymol by an undergrad working in the Structural Biology lab at UC Davis.

Protein Crystallisation

To perform protein crystallography, a reliable source of protein must be available, together with a purification/concentration protocol that will yield high quality, homogeneous, soluble material.

The growth of protein crystals of sufficient quality for structure determination is, without doubt, the rate limiting step in most protein crystallographic work, and is the least well understood. The principle of crystallisation, whether of macromolecules or salts (unfortunately!) is to take a solution of the sample at high concentration and induce it to come out of solution; if this happens too fast then precipitation will occur, but under the correct conditions crystals will grow. The elucidation of these conditions determines the rate limiting step and indeed whether or not the project will be possible. Many projects prove not to be possible because of the inability to crystallise the protein. The magnitude of the problem can be understood when one considers the variables the choice of precipitant, its concentration, the buffer, its pH, the protein concentration, the temperature, the crystallisation technique, and the possible inclusion of additives. Essentially, initial experiments will be based on a trial and error procedure, which aims to cover as wide a range as possible of the variables as is practical. Commercially available "crystal screen" packages are often used at this stage. Each one usually consists of 50 solutions varying widely in precipitant, buffer, pH, and salt, known as a sparse matrix. These can then be set up using the techniques of sitting drop vapour diffusion, hanging drop vapour diffusion, and possibly dialysis,usually at both room temperature and 4°C. In this way, many of the variables can be covered easily, and one or more might even yield crystals of sufficient quality to proceed to the next step. However, it is more usual at this stage to see various combinations of the following: nothing, precipitation, showers of microcrystals (which often resemble a precipitate), or a few very tiny crystals. Obtaining either of the last two results is encouraging because it indicates that the macromolecular subunit might possess sufficient inherent structural order or symmetry to make crystallisation possible. In general, those proteins that are glycosylated or contain flexible or less conformationally constrained domains are difficult to crystallise, whereas even very large complexes of high symmetry, such as many viruses, will crystallise. Various techniques can be used to improve crystal size. These include seeding, alteration of protein concentration, or alteration of temperature. For diffraction analysis, protein crystals are usually required to be a minimum of 0.1 mm in the longest dimension, to provide a sufficient volume of crystal lattice that can be exposed to the beam.

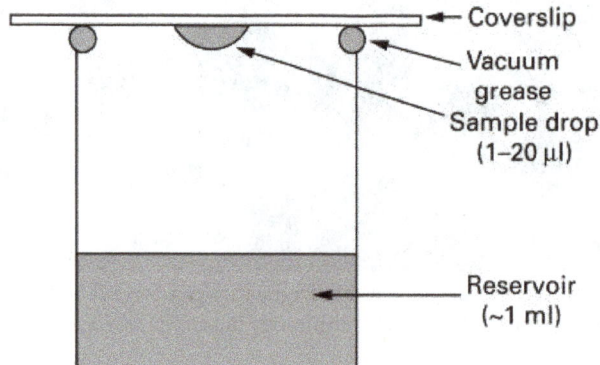

Hanging drop vapour diffusion is usually set up with a drop of 1–20 µl suspended from a glass coverslip over the reservoir solution. The drop is a 50/50 (vol/vol) mixture of the protein solution and the reservoir solution. Hence, the vapour pressure of water around the drop is greater than that over the reservoir. The pressure gradient across the vapour space leads to a net loss of water in the drop. Sitting drop vapour diffusion differs only in that the drop sits in a depression on a specially constructed raised platform within the diffusion chamber.

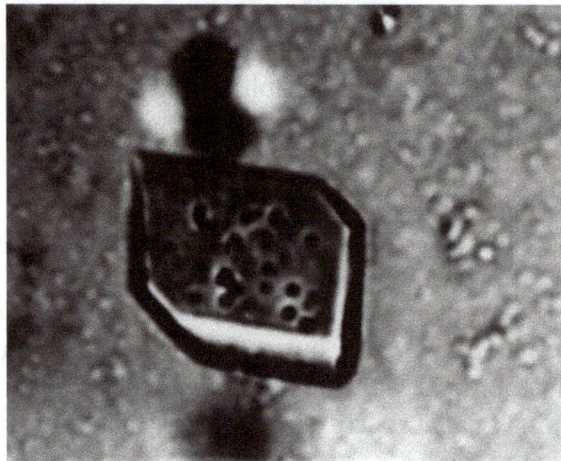

A crystal of a bovine picornavirus measuring 0.2 mm in the longest dimension grown, using ammonium sulphate as precipitant and using microdialysis apparatus. These crystals take 48 hours to grow at room temperature, but are very fragile because the virus particles in the crystal are roughly spherical and 300 Å in diameter, making the solvent content very high and the particle contacts very low. This is common with many protein crystals and great care must be taken during manipulation.

Before proceeding, it must be verified that the crystals contain the desired macromolecule as opposed to one of the salts from the precipitant or buffer. This can be done by sacrificing some crystals to polyacrylamide gel electrophoresis analysis, staining, or a test xray diffraction exposure.

Data Analysis

Crystal Symmetry, Unit Cell, and Image Scaling

The recorded series of two-dimensional diffraction patterns, each corresponding to a different crystal orientation, is converted into a three-dimensional model of the electron density; the conversion uses the mathematical technique of Fourier transforms, which is explained below. Each spot corresponds to a different type of variation in the electron density; the crystallographer must determine *which* variation corresponds to *which* spot (*indexing*), the relative strengths of the spots in different images (*merging and scaling*) and how the variations should be combined to yield the total electron density (*phasing*).

Data processing begins with *indexing* the reflections. This means identifying the dimensions of the unit cell and which image peak corresponds to which position in reciprocal space. A byproduct of indexing is to determine the symmetry of the crystal, i.e., its *space group*. Some space groups can be eliminated from the beginning. For example, reflection symmetries cannot be observed in chiral molecules; thus, only 65 space groups of 230 possible are allowed for protein molecules which are almost always chiral. Indexing is generally accomplished using an *autoindexing* routine. Having assigned symmetry, the data is then *integrated*. This converts the hundreds of images containing the thousands of reflections into a single file, consisting of (at the very least) records of the Miller index of each reflection, and an intensity for each reflection (at this state the file often also includes error estimates and measures of partiality (what part of a given reflection was recorded on that image)).

A full data set may consist of hundreds of separate images taken at different orientations of the crystal. The first step is to merge and scale these various images, that is, to identify which peaks appear in two or more images (*merging*) and to scale the relative images so that they have a consistent intensity scale. Optimizing the intensity scale is critical because the relative intensity of the peaks is the key information from which the structure is determined. The repetitive technique of crystallographic data collection and the often high symmetry of crystalline materials cause the diffractometer to record many symmetry-equivalent reflections multiple times. This allows calculating the symmetry-related R-factor, a reliability index based upon how similar are the measured intensities of symmetry-equivalent reflections, thus assessing the quality of the data.

Initial Phasing

The data collected from a diffraction experiment is a reciprocal space representation of the crystal lattice. The position of each diffraction 'spot' is governed by the size and shape of the unit cell, and the inherent symmetry within the crystal. The intensity of each diffraction 'spot' is recorded, and this intensity is proportional to the square of the *structure factor* amplitude. The structure factor is a complex number containing information relating to both the amplitude and phase of a wave. In order to obtain an interpretable *electron density map*, both amplitude and phase must be known (an electron density map allows a crystallographer to build a starting model of the molecule). The phase cannot be directly recorded during a diffraction experiment: this is known as the phase problem. Initial phase estimates can be obtained in a variety of ways:

- *Ab initio* phasing or direct methods – This is usually the method of choice for small molecules (<1000 non-hydrogen atoms), and has been used successfully to solve the phase problems for small proteins. If the resolution of the data is better than 1.4 Å (140 pm), direct methods can be used to obtain phase information, by exploiting known phase relationships between certain groups of reflections.

- Molecular replacement – if a related structure is known, it can be used as a search model in molecular replacement to determine the orientation and position of the molecules within the unit cell. The phases obtained this way can be used to generate *electron density maps*.

- Anomalous X-ray scattering (*MAD or SAD phasing*) – the X-ray wavelength may be scanned past an absorption edge of an atom, which changes the scattering in a known way. By recording full sets of reflections at three different wavelengths (far below, far above and in the middle of the absorption edge) one can solve for the substructure of the anomalously diffracting atoms and hence the structure of the whole molecule. The most popular method of incorporating anomalous scattering atoms into proteins is to express the protein in a methionine auxotroph (a host incapable of synthesizing methionine) in a media rich in seleno-methionine, which contains selenium atoms. A MAD experiment can then be conducted around the absorption edge, which should then yield the position of any methionine residues within the protein, providing initial phases.

- Heavy atom methods (multiple isomorphous replacement) – If electron-dense metal atoms can be introduced into the crystal, direct methods or Patterson-space methods can be used to determine their location and to obtain initial phases. Such heavy atoms can be introduced either by soaking the crystal in a heavy atom-containing solution, or by co-crystallization (growing the crystals in the presence of a heavy atom). As in MAD phasing, the changes in the scattering amplitudes can be interpreted to yield the phases. Although this is the original method by which protein crystal structures were solved, it has largely been superseded by MAD phasing with selenomethionine.

Model Building and Phase Refinement

Having obtained initial phases, an initial model can be built. This model can be used to refine the phases, leading to an improved model, and so on. Given a model of some atomic positions, these positions and their respective Debye-Waller factors (or B-factors, accounting for the thermal motion of the atom) can be refined to fit the observed diffraction data, ideally yielding a better set of phases. A new model can then be fit to the new electron density map and a further round of refinement is carried out. This continues until the correlation between the diffraction data and the model is maximized. The agreement is measured by an *R*-factor defined as

$$R = \frac{\sum_{\text{all reflections}} |F_o - F_c|}{\sum_{\text{all reflections}} |F_o|}$$

where F is the structure factor. A similar quality criterion is R_{free}, which is calculated from a subset (~10%) of reflections that were not included in the structure refinement. Both R factors depend on the resolution of the data. As a rule of thumb, R_{free} should be approximately the resolution in angstroms divided by 10; thus, a data-set with 2 Å resolution should yield a final R_{free} ~ 0.2. Chemical bonding features such as stereochemistry, hydrogen bonding and distribution of bond lengths and angles are complementary measures of the model quality. Phase bias is a serious problem in such iterative model building. *Omit maps* are a common technique used to check for this.

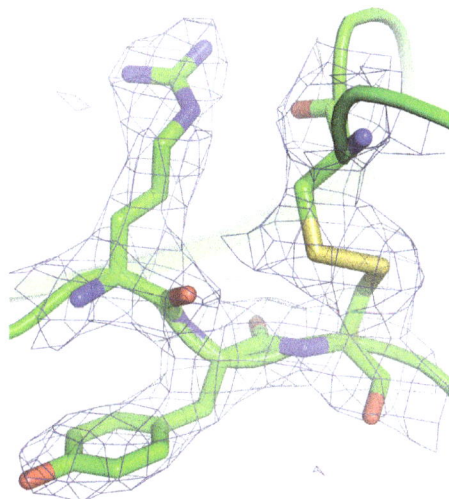

A protein crystal structure at 2.7 Å resolution. The mesh encloses the region in which the electron density exceeds a given threshold.

The straight segments represent chemical bonds between the non-hydrogen atoms of an arginine (upper left), a tyrosine (lower left), a disulfide bond (upper right, in yellow), and some peptide groups (running left-right in the middle). The two curved green tubes represent spline fits to the polypeptide backbone.

It may not be possible to observe every atom in the asymmetric unit. In many cases, Crystallographic disorder smears the electron density map. Weakly scattering atoms such as hydrogen are routinely invisible. It is also possible for a single atom to appear multiple times in an electron density map, e.g., if a protein sidechain has multiple (<4) allowed conformations. In still other cases, the crystallographer may detect that the covalent structure deduced for the molecule was incorrect, or changed. For example, proteins may be cleaved or undergo post-translational modifications that were not detected prior to the crystallization.

Disorder

A common challenge in refinement of crystal structures results from crystallographic disorder. Disorder can take many forms but in general involves the coexistence of two or more species or conformations. Failure to recognize disorder results in flawed interpretation. Pitfalls from improper modeling of disorder are illustrated by the discounted hypothesis of bond stretch isomerism. Disorder is modelled with respect to the relative population of the components, often only two, and their identity. In structures of large molecules and ions, solvent and counterions are often disordered.

Deposition of the Structure

Once the model of a molecule's structure has been finalized, it is often deposited in a crystallographic database such as the Cambridge Structural Database (for small molecules), the Inorganic Crystal Structure Database (ICSD) (for inorganic compounds) or the Protein Data Bank (for protein structures). Many structures obtained in private commercial ventures to crystallize medicinally relevant proteins are not deposited in public crystallographic databases.

Contributions to Chemistry and Material Science

X-ray crystallography has led to a better understanding of chemical bonds and non-covalent interactions. The initial studies revealed the typical radii of atoms, and confirmed many theoretical models of chemical bonding, such as the tetrahedral bonding of carbon in the diamond structure, the octahedral bonding of metals observed in ammonium hexachloroplatinate (IV), and the resonance observed in the planar carbonate group and in aromatic molecules. Kathleen Lonsdale's 1928 structure of hexamethylbenzene established the hexagonal symmetry of benzene and showed a clear difference in bond length between the aliphatic C–C bonds and aromatic C–C bonds; this finding led to the idea of resonance between chemical bonds, which had profound consequences for the development of chemistry. Her conclusions were anticipated by William Henry Bragg, who published models of naphthalene and anthracene in 1921 based on other molecules, an early form of molecular replacement.

Also in the 1920s, Victor Moritz Goldschmidt and later Linus Pauling developed rules for eliminating chemically unlikely structures and for determining the relative sizes of atoms. These rules led to the structure of brookite (1928) and an understanding of the relative stability of the rutile, brookite and anatase forms of titanium dioxide.

The distance between two bonded atoms is a sensitive measure of the bond strength and its bond order; thus, X-ray crystallographic studies have led to the discovery of even more exotic types of bonding in inorganic chemistry, such as metal-metal double bonds, metal-metal quadruple bonds, and three-center, two-electron bonds. X-ray crystallography—or, strictly speaking, an inelastic Compton scattering experiment—has also provided evidence for the partly covalent character of hydrogen bonds. In the field of organometallic chemistry, the X-ray structure of ferrocene initiated scientific studies of sandwich compounds, while that of Zeise's salt stimulated research into "back bonding" and metal-pi complexes. Finally, X-ray crystallography had a pioneering role in the development of supramolecular chemistry, particularly in clarifying the structures of the crown ethers and the principles of host-guest chemistry.

In material sciences, many complicated inorganic and organometallic systems have been analyzed using single-crystal methods, such as fullerenes, metalloporphyrins, and other complicated compounds. Single-crystal diffraction is also used in the pharmaceutical industry, due to recent problems with polymorphs. The major factors affecting the quality of single-crystal structures are the crystal's size and regularity; recrystallization is a commonly used technique to improve these factors in small-molecule crystals. The Cambridge Structural Database contains over 800,000 structures as of September 2016; over 99% of these structures were determined by X-ray diffraction.

Mineralogy and Metallurgy

Since the 1920s, X-ray diffraction has been the principal method for determining the arrangement of atoms in minerals and metals. The application of X-ray crystallography to mineralogy began with the structure of garnet, which was determined in 1924 by Menzer. A systematic X-ray crystallographic study of the silicates was undertaken in the 1920s. This study showed that, as the Si/O ratio is altered, the silicate crystals exhibit significant changes in their atomic arrangements. Machatschki extended these insights to minerals in which aluminium substitutes for the silicon atoms of the silicates. The first application of X-ray crystallography to metallurgy likewise occurred in the mid-1920s. Most notably, Linus Pauling's structure of the alloy Mg_2Sn led to his theory of the stability and structure of complex ionic crystals.

First X-ray diffraction view of Martian soil – CheMin analysis reveals feldspar, pyroxenes, olivine and more (Curiosity rover at "Rocknest", October 17, 2012).

On October 17, 2012, the Curiosity rover on the planet Mars at "Rocknest" performed the first X-ray diffraction analysis of Martian soil. The results from the rover's CheMin analyzer revealed the presence of several minerals, including feldspar, pyroxenes and olivine, and suggested that the Martian soil in the sample was similar to the "weathered basaltic soils" of Hawaiian volcanoes.

Early Organic and small Biological Molecules

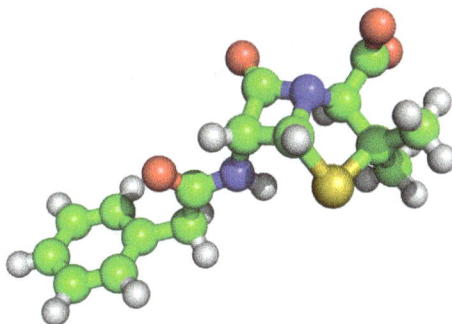

The three-dimensional structure of penicillin, solved by Dorothy Crowfoot Hodgkin in 1945. The green, white, red, yellow and blue spheres represent atoms of carbon, hydrogen, oxygen, sulfur and nitrogen, respectively.

The first structure of an organic compound, hexamethylenetetramine, was solved in 1923. This was followed by several studies of long-chain fatty acids, which are an important component of biological membranes. In the 1930s, the structures of much larger molecules with two-dimensional complexity began to be solved. A significant advance was the structure of phthalocyanine, a large planar molecule that is closely related to porphyrin molecules important in biology, such as heme, corrin and chlorophyll.

X-ray crystallography of biological molecules took off with Dorothy Crowfoot Hodgkin, who solved the structures of cholesterol (1937), penicillin (1946) and vitamin B_{12} (1956), for which she was awarded the Nobel Prize in Chemistry in 1964. In 1969, she succeeded in solving the structure of insulin, on which she worked for over thirty years.

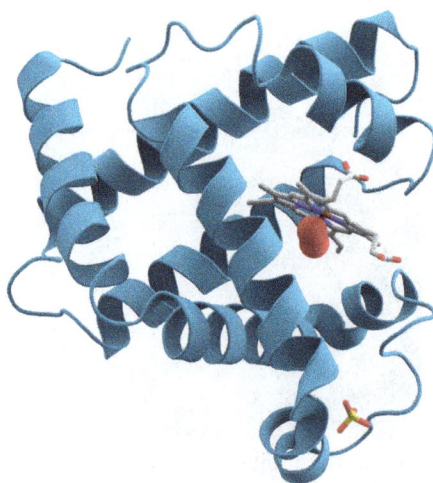

Ribbon diagram of the structure of myoglobin, showing colored alpha helices.

Such proteins are long, linear molecules with thousands of atoms; yet the relative position of each atom has been determined with sub-atomic resolution by X-ray crystallography. Since it is difficult to visualize all the atoms at once, the ribbon shows the rough path of the protein polymer from its N-terminus (blue) to its C-terminus (red).

Biological Macromolecular Crystallography

Crystal structures of proteins (which are irregular and hundreds of times larger than cholesterol) began to be solved in the late 1950s, beginning with the structure of sperm whale myoglobin by Sir John Cowdery Kendrew, for which he shared the Nobel Prize in Chemistry with Max Perutz in 1962. Since that success, 132055 X-ray crystal structures of proteins, nucleic acids and other biological molecules have been determined. For comparison, the nearest competing method in terms of structures analyzed is nuclear magnetic resonance (NMR) spectroscopy, which has resolved 11904 chemical structures. Moreover, crystallography can solve structures of arbitrarily large molecules, whereas solution-state NMR is restricted to relatively small ones (less than 70 kDa). X-ray crystallography is now used routinely by scientists to determine how a pharmaceutical drug interacts with its protein target and what changes might improve it. However, intrinsic membrane proteins remain challenging to crystallize because they require detergents or other means to solubilize them in isolation, and such detergents often interfere with crystallization. Such membrane proteins are a large component of the genome, and include many proteins of great physiological importance, such as ion channels and receptors. Helium cryogenics are used to prevent radiation damage in protein crystals.

On the other end of the size scale, even relatively small molecules may pose challenges for the resolving power of X-ray crystallography. The structure assigned in 1991 to the antibiotic isolated from a marine organism, diazonamide A ($C_{40}H_{34}Cl_2N_6O_6$, molar mass 765.65 g/mol), proved to be incorrect by the classical proof of structure: a synthetic sample was not identical to the natural product. The mistake was attributed to the inability of X-ray crystallography to distinguish between the correct -OH / >NH and the interchanged -NH$_2$ / -O- groups in the incorrect structure. With advances in instrumentation, however, it is now routinely possible to distinguish between such similar groups using modern single-crystal X-ray diffractometers.

Relationship to other Scattering Techniques

Elastic vs. Inelastic Scattering

X-ray crystallography is a form of elastic scattering; the outgoing X-rays have the same energy, and thus same wavelength, as the incoming X-rays, only with altered direction. By contrast, *inelastic scattering* occurs when energy is transferred from the incoming X-ray to the crystal, e.g., by exciting an inner-shell electron to a higher energy level. Such inelastic scattering reduces the energy (or increases the wavelength) of the outgoing beam. Inelastic scattering is useful for probing such excitations of matter, but not in determining the distribution of scatterers within the matter, which is the goal of X-ray crystallography.

X-rays range in wavelength from 10 to 0.01 nanometers; a typical wavelength used for crystallography is 1 Å (0.1 nm), which is on the scale of covalent chemical bonds and the radius of a single atom. Longer-wavelength photons (such as ultraviolet radiation) would not have sufficient resolution to determine the atomic positions. At the other extreme, shorter-wavelength photons such as gamma rays are difficult to produce in large numbers, difficult to focus, and interact too strongly with matter, producing particle-antiparticle pairs. Therefore, X-rays are the "sweetspot" for wavelength when determining atomic-resolution structures from the scattering of electromagnetic radiation.

Other X-ray Techniques

Other forms of elastic X-ray scattering include powder diffraction, Small-Angle X-ray Scattering (SAXS) and several types of X-ray fiber diffraction, which was used by Rosalind Franklin in determining the double-helix structure of DNA. In general, single-crystal X-ray diffraction offers more structural information than these other techniques; however, it requires a sufficiently large and regular crystal, which is not always available.

These scattering methods generally use *monochromatic* X-rays, which are restricted to a single wavelength with minor deviations. A broad spectrum of X-rays (that is, a blend of X-rays with different wavelengths) can also be used to carry out X-ray diffraction, a technique known as the Laue method. This is the method used in the original discovery of X-ray diffraction. Laue scattering provides much structural information with only a short exposure to the X-ray beam, and is therefore used in structural studies of very rapid events (Time resolved crystallography). However, it is not as well-suited as monochromatic scattering for determining the full atomic structure of a crystal and therefore works better with crystals with relatively simple atomic arrangements.

The Laue back reflection mode records X-rays scattered backwards from a broad spectrum source. This is useful if the sample is too thick for X-rays to transmit through it. The diffracting planes in the crystal are determined by knowing that the normal to the diffracting plane bisects the angle between the incident beam and the diffracted beam. A Greninger chart can be used to interpret the back reflection Laue photograph.

Electron and Neutron Diffraction

Other particles, such as electrons and neutrons, may be used to produce a diffraction pattern. Although electron, neutron, and X-ray scattering are based on different physical processes, the resulting diffraction patterns are analyzed using the same coherent diffraction imaging techniques.

As derived below, the electron density within the crystal and the diffraction patterns are related by a simple mathematical method, the Fourier transform, which allows the density to be calculated relatively easily from the patterns. However, this works only if the scattering is *weak*, i.e., if the scattered beams are much less intense than the incoming beam. Weakly scattered beams pass through the remainder of the crystal without undergoing a second scattering event. Such re-scattered waves are called "secondary scattering" and hinder the analysis. Any sufficiently thick crystal will produce secondary scattering, but since X-rays interact relatively weakly with the electrons, this is generally not a significant concern. By contrast, electron beams may produce strong secondary scattering even for relatively thin crystals (>100 nm). Since this thickness corresponds to the diameter of many viruses, a promising direction is the electron diffraction of isolated macromolecular assemblies, such as viral capsids and molecular machines, which may be carried out with a cryo-electron microscope. Moreover, the strong interaction of electrons with matter (about 1000 times stronger than for X-rays) allows determination of the atomic structure of extremely small volumes. The field of applications for electron crystallography ranges from bio molecules like membrane proteins over organic thin films to the complex structures of (nanocrystalline) intermetallic compounds and zeolites.

Neutron diffraction is an excellent method for structure determination, although it has been difficult to obtain intense, monochromatic beams of neutrons in sufficient quantities. Traditionally, nuclear reactors have been used, although sources producing neutrons by spallation are becoming increasingly available. Being uncharged, neutrons scatter much more readily from the atomic nuclei rather than from the electrons. Therefore, neutron scattering is very useful for observing the positions of light atoms with few electrons, especially hydrogen, which is essentially invisible in the X-ray diffraction. Neutron scattering also has the remarkable property that the solvent can be made invisible by adjusting the ratio of normal water, H_2O, and heavy water, D_2O.

Electron Crystallography

Comparison with X-ray crystallography

It can complement X-ray crystallography for studies of very small crystals (<0.1 micrometers), both inorganic, organic, and proteins, such as membrane proteins, that cannot easily form the large 3-dimensional crystals required for that process. Protein structures are usually determined from either

2-dimensional crystals (sheets or helices), polyhedrons such as viral capsids, or dispersed individual proteins. Electrons can be used in these situations, whereas X-rays cannot, because electrons interact more strongly with atoms than X-rays do. Thus, X-rays will travel through a thin 2-dimensional crystal without diffracting significantly, whereas electrons can be used to form an image. Conversely, the strong interaction between electrons and protons makes thick (e.g. 3-dimensional > 1 micrometer) crystals impervious to electrons, which only penetrate short distances.

One of the main difficulties in X-ray crystallography is determining phases in the diffraction pattern. Because of the complexity of X-ray lenses, it is difficult to form an image of the crystal being diffracted, and hence phase information is lost. Fortunately, electron microscopes can resolve atomic structure in real space and the crystallographic structure factor phase information can be experimentally determined from an image's Fourier transform. The Fourier transform of an atomic resolution image is similar, but different, to a diffraction pattern—with reciprocal lattice spots reflecting the symmetry and spacing of a crystal. Aaron Klug was the first to realize that the phase information could be read out directly from the Fourier transform of an electron microscopy image that had been scanned into a computer, already in 1968. For this, and his studies on virus structures and transfer-RNA, Klug received the Nobel Prize for chemistry in 1982.

Radiation Damage

A common problem to X-ray crystallography and electron crystallography is radiation damage, by which especially organic molecules and proteins are damaged as they are being imaged, limiting the resolution that can be obtained. This is especially troublesome in the setting of electron crystallography, where that radiation damage is focused on far fewer atoms. One technique used to limit radiation damage is electron cryomicroscopy, in which the samples undergo cryofixation and imaging takes place at liquid nitrogen or even liquid helium temperatures. Because of this problem, X-ray crystallography has been much more successful in determining the structure of proteins that are especially vulnerable to radiation damage.

Protein Structures Determined by Electron Crystallography

The first electron crystallographic protein structure to achieve atomic resolution was bacteriorhodopsin, determined by Richard Henderson and coworkers at the Medical Research Council Laboratory of Molecular Biology in 1990. However, already in 1975 Unwin and Henderson had determined the first membrane protein structure at intermediate resolution (7 Ångström), showing for the first time the internal structure of a membrane protein, with its alpha-helices standing perpendicular to the plane of the membrane. Since then, several other high-resolution structures have been determined by electron crystallography, including the light-harvesting complex, the nicotinic acetylcholine receptor, and the bacterial flagellum.

Application to Inorganic Materials

Electron crystallographic studies on inorganic crystals using high-resolution electron microscopy (HREM) images were first performed by Aaron Klug in 1978 and by Sven Hovmöller and coworkers in 1984. HREM images were used because they allow to select (by computer software) only the very thin regions close to the edge of the crystal for structure analysis. This is of crucial importance since in the thicker parts of the crystal the exit-wave function (which carries the information about the intensity and position of the projected atom columns) is no longer linearly related to the projected crystal structure. Moreover, not only do the HREM images change their appearance with increasing crystal thickness, they are also very sensitive to the chosen setting of the defocus Δf of the objective lens. To cope with this complexity Michael O'Keefe started in the early 1970s to develop image simulation software which allowed to understand an interpret the observed contrast changes in HREM images.

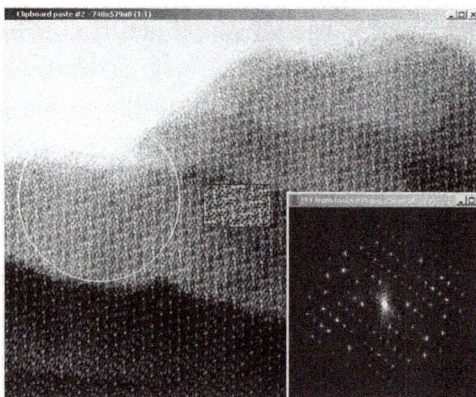

Electron microscopy image of an inorganic tantalum oxide, with its Fourier transform, inset.

Notice how the appearance changes from the upper thin region to the thicker lower region. The unit cell of this compound is about 15 by 25 Ångström. It is outlined at the centre of the figure, inside the result from image processing, where the symmetry has been taken into account. The black dots show clearly all the tantalum atoms. The diffraction extends to 6 orders along the 15 Å direction and 10 orders in the perpendicular direction. Thus the resolution of the EM image is 2.5 Å (15/6 or 25/10). This calculated Fourier transform contain both amplitudes (as seen) and phases (not displayed).

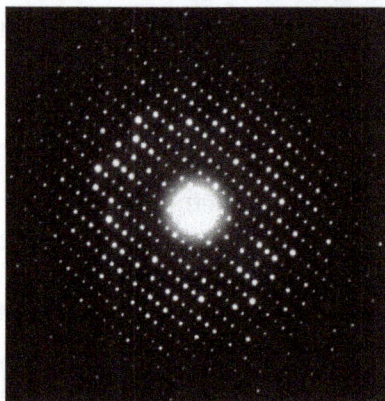

Electron diffraction pattern of the same crystal of inorganic tantalum oxide shown above.

Notice that there are many more diffraction spots here than in the diffractogram calculated from the EM image above. The diffraction extends to 12 orders along the 15 Å direction and 20 orders in the perpendicular direction. Thus the resolution of the ED pattern is 1.25 Å (15/12 or 25/20). ED patterns do not contain phase information, but the clear differences between intensities of the diffraction spots can be used in crystal structure determination.

There was a serious disagreement in the field of electron microscopy of inorganic compounds;

while some have claimed that "the phase information is present in EM images" others have the opposite view that "the phase information is lost in EM images". The reason for these opposite views is that the word "phase" has been used with different meanings in the two communities of physicists and crystallographers. The physicists are more concerned about the "electron wave phase" - the phase of a wave moving through the sample during exposure by the electrons. This wave has a wavelength of about 0.02-0.03 Ångström (depending on the accelerating voltage of the electron microscope). Its phase is related to the phase of the undiffracted direct electron beam. The crystallographers, on the other hand, mean the "crystallographic structure factor phase" when they simply say "phase". This phase is the phase of standing waves of potential in the crystal (very similar to the electron density measured in X-ray crystallography). Each of these waves have their specific wavelength, called d-value for distance between so-called Bragg planes of low/high potential. These d-values range from the unit cell dimensions to the resolution limit of the electron microscope, i.e. typically from 10 or 20 Ångströms down to 1 or 2 Ångströms. Their phases are related to a fixed point in the crystal, defined in relation to the symmetry elements of that crystal. The crystallographic phases are a property of the crystal, so they exist also outside the electron microscope. The electron waves vanish if the microscope is switched off. In order to determine a crystal structure, it is necessary to know the crystallographic structure factors, but not to know the electron wave phases. A more detailed discussion how (crystallographic structure factor) phases link with the phases of the electron wave can be found in.

Just as with proteins, it has been possible to determine the atomic structures of inorganic crystals by electron crystallography. For simpler structure it is sufficient to use three perpendicular views, but for more complicated structures, also projections down ten or more different diagonals may be needed.

In addition to electron microscopy images, it is also possible to use electron diffraction (ED) patterns for crystal structure determination. The utmost care must be taken to record such ED patterns from the thinnest areas in order to keep most of the structure related intensity differences between the reflections (quasi-kinematical diffraction conditions). Just as with X-ray diffraction patterns, the important crystallographic structure factor phases are lost in electron diffraction patterns and must be uncovered by special crystallographic methods such as direct methods, maximum likelihood or (more recently) by the charge-flipping method. On the other hand, ED patterns of inorganic crystals have often a high resolution (= interplanar spacings with high Miller indices) much below 1 Ångström. This is comparable to the point resolution of the best electron microscopes. Under favourable conditions it is possible to use ED patterns from a single orientation to determine the complete crystal structure. Alternatively a hybrid approach can be used which uses HRTEM images for solving and intensities from ED for refining the crystal structure.

Recent progress for structure analysis by ED was made by introducing the Vincent-Midgley precession technique for recording electron diffraction patterns. The thereby obtained intensities are usually much closer to the kinematical intensities, so that even structures can be determined that are out of range when processing conventional (selected area) electron diffraction data.

Crystal structures determined via electron crystallography can be checked for their quality by using first-principles calculations within density functional theory (DFT). This approach was for the first time applied for the validation of several metal-rich structures which were only accessible by HRTEM and ED, respectively.

Recently, two very complicated zeolite structures have been determined by electron crystallography combined with X-ray powder diffraction. These are more complex than the most complex zeolite structures determined by X-ray crystallography.

Nuclear Magnetic Resonance Crystallography

Nuclear Magnetic Resonance (NMR) crystallography is a type method that uses primary NMR spectroscopy to find the structure of different solid materials in the atomic scale. So the solid-state NMR spectroscopy will be used primarily, and possibly supplemented by quantum chemistry calculations (e.g. density functional theory), powder diffraction etc. If crystals is grown is properly and uniquely, any crystallographic method can generally be used to determine the crystal structure and in case of organic compounds the molecular structures and molecular packing. The main use of NMR crystallography is in determining micro crystalline materials which are used to this method but not to X-ray, neutron and electron diffraction. This is largely used because interactions that are short range are measured in NMR crystallography.

- Dipolar interaction

- Non-covalent interactions

- Solid-State NMR

- Crystal Structure Refinements

- Chemical shift interaction

The term 'NMR Crystallography' presents a broad polysemy. To some, it represents a stand-alone structure elucidation method for single crystal, polycrystalline or amorphous compounds. For others, it is a source of additional structural information when compounds fail to yield crystals of sufficient quality or size suitable for single-crystal diffraction-based structure determination, or when powder diffraction patterns exhibit a too high degree of complexity for structure model elaboration. Some consider NMR, diffraction and modelling as a synergistic complementary set of methods. Others consider that the multiplicity of specific NMR experiments allows for the progressive build-up of topological sub-graphs of the crystal graph, and thus drives the structure model search. These are all established uses of magnetic resonance toward the investigation of the crystalline state.

When applied to organic molecules, NMR crystallography aims at including structural information not only of a single molecule but also on the molecular packing (i.e. crystal structure). Contrary to X-ray, single crystals are not necessary with solid-state NMR and structural information can be obtained from high-resolution spectra of disordered solids. E.g. polymorphism is an area of interest for NMR crystallography since this is encountered occasionally (and may often be previously undiscovered) in organic compounds. In this case a change in the molecular structure and/or in the molecular packing can lead to polymorphism, and this can be investigated by NMR crystallography.

Dipolar Couplings-based Approach

The spin interaction that is usually employed for structural analyses via solid state NMR spectroscopy is the magnetic dipolar interaction. Additional knowledge about other interactions within the studied system like the chemical shift or the electric quadrupole interaction can be helpful as well, and in some cases solely the chemical shift has been employed as e.g. for zeolites. The "dipole coupling"-based approach parallels protein NMR spectroscopy to some extent in that e.g. multiple residual dipolar couplings are measured for proteins in solution, and these couplings are used as constraints in the protein structure calculation.

In NMR crystallography the observed spins in case of organic molecules would often be spin-1/2 nuclei of moderate frequency ($^{13}C, ^{15}N, ^{31}P$, etc.). I.e. ^{1}H is excluded due to its large magnetogyric ratio and high spin concentration leading to a network of strong homonuclear dipolar couplings. There are two solutions with respect to $^{1}H{:}^{1}H$ spin diffusion experiments and specific labelling with 2H spins (spin = 1). The latter is also popular e.g. in NMR spectroscopic investigations of hydrogen bonds in solution and the solid state.[10] Both intra- and intermolecular structural elements can be investigated e.g. via deuterium REDOR (an established solid state NMR pulse sequence to measure dipolar couplings between deuterons and other spins). This can provide an additional constraint for an NMR crystallographic structural investigation in that it can be used to find and characterize e.g. intermolecular hydrogen bonds.

Dipolar Interaction

The above-mentioned dipolar interaction can be measured directly, e.g. between pairs of heteronuclear spins like $^{13}C/^{15}N$ in many organic compounds. Furthermore, the strength of the dipolar interaction modulates parameters like the longitudinal relaxation time or the spin diffusion rate which therefore can be examined to obtain structural information. E.g. ^{1}H spin diffusion has been measured providing rich structural information.

Chemical Shift Interaction

The chemical shift interaction can be used in conjunction with the dipolar interaction to determine the orientation of the dipolar interaction frame (principal axes system) with respect to the molecular frame (dipolar chemical shift spectroscopy). For some cases there are rules for the chemical shift interaction tensor orientation as for the ^{13}C spin in ketones due to symmetry arguments (sp^2 hybridisation). If the orientation of a dipolar interaction (between the spin of interest and e.g. another heteronucleus) is measured with respect to the chemical shift interaction coordinate system, these two pieces of information (chemical shift tensor/molecular orientation and the dipole tensor/chemical shift tensor orientation) combined give the orientation of the dipole tensor in the molecular frame. However, this method is only suitable for small molecules (or polymers with a small repetition unit like polyglycine) and it provides only selective (and usually intramolecular) structural information.

Crystal Structure Refinements

The dipolar interaction yields the most direct information with respect to structure as it makes it possible to measure the distances between the spins. The sensitivity of this interaction is however lacking and even though dipolar-based NMR crystallography makes the elucidation of structures possible, other methods are necessary to obtain high resolution structures. For these reasons much work was done to include the use other NMR observables such as chemical shift anisotropy,

J-coupling and the quadrupolar interaction. These anisotropic interactions are highly sensitive to the 3D local environment making it possible to refine the structures of powdered samples to structures rivaling the quality of single crystal X-ray diffraction. These however rely on adequate methods for predicting these interactions as they do not depend in a straightforward fashion on the structure.

Comparison with Diffraction Methods

A drawback of NMR crystallography is that the method is typically more time consuming and more expensive (due to spectrometer costs and isotope labelling) than X-ray crystallography, it often elucidates only part of the structure, and isotope labelling and experiments may have to be tailored to obtain key structural information. Also not always is a molecular structure suitable for a pure NMR-based NMR crystallographic approach, but it can still play an important role in a multimodality (NMR+diffraction) study.

Unlike in the case of diffraction methods, it appears that NMR crystallography needs to work on a case by case basis. This is the case since difference systems will have different spin physics and different observables which can be probed. The method may therefore not find widespread use as different systems will require qualified individuals to design experiments to study them.

R-factor

The term R factor in crystallography is commonly taken to refer to the 'conventional' R factor

$$R = \frac{\sum |F_{obs} - F_{calc}|}{\sum |F_{obs}|}$$

a measure of agreement between the amplitudes of the structure factors calculated from a crystallographic model and those from the original X-ray diffraction data. The R factor is calculated during each cycle of least-squares structure refinement to assess progress. The final R factor is one measure of model quality.

More generally, a variety of R factors may be determined to measure analogous residuals during least-squares optimization procedures. Where the refinement attempts to minimize the deviates of the squares of the structure factors (refinement based on F^2), the R factor based on F^2 is used to monitor the progress of refinement:

$$R(F^2) = \frac{\sum |F_{obs}^2 - F_{calc}^2|}{\sum |F_{obs}^2|}$$

Likewise, refinement based on I can be tracked using the Bragg R factor

$$R_B = \frac{\sum |I_{obs} - I_{calc}|}{\sum |I_{obs}|}$$

Even for refinement based on F2 or I, the 'conventional' R factor may be calculated and quoted as a measure of model quality, in order to compare the resulting quality of models calculated at different times and with different refinement strategies.

The R factor is sometimes described as the discrepancy index.

R factor as a Measure of Structure Quality

Theoretical values of R range from zero (perfect agreement of calculated and observed intensities) to about 0.6 for a set of measured intensities compared against a set of random intensities. R factors greater than 0.5 indicate very poor agreement between observed and calculated intensities, and many models with $R \geq 0.5$ will not respond to attempts at improvement. An early model with $R \leq 0.4$ can usually be improved during refinement. A desirable target R factor for a protein model refined with data to 2.5 Å is considered to be ~ 0.2. Small organic molecules commonly refine to R < 0.05. However, the R factor must always be treated with caution, as an indicator of precision and not accuracy. Partially incorrect structures have been reported with R values below 0.1; many imprecise but essentially correct structures have been reported with higher R values.

Weighted R Factors

In practice, weighted R factors are more often used to track least-squares refinement, since the functions minimized are weighted according to estimates of the precision of the measured quantity y: $\sum w(Y_o - Y_c)^2$ (Y being F, F² or I). The general term for a weighted residual is

$$wR = \left(\frac{\sum |w| Y_o - Y_c|^2|}{\sum |wY_o^2|} \right)^{1/2}$$

The sum is usually computed over all reflections measured in the experiment. However, occasionally reflections are omitted from the calculation, either because they are believed to result from a systematic experimental error or are recorded with an intensity small compared with background noise. Any such selection may introduce statistical artefacts, and must always be described when reporting R factors.

References

- Shafranovskii I I & Belov N V (1962). Paul Ewald, ed. "E. S. Fedorov" (PDF). 50 Years of X-Ray Diffraction. Springer: 351. ISBN 90-277-9029-9

- "Cryogenic (<20 K) helium cooling mitigates radiation damage to protein crystals". Acta Crystallographica D. 63 (4): 486–492. 2007. doi:10.1107/s0907444907005264

- von Laue M (1914). "Concerning the detection of x-ray interferences" (PDF). Nobel Lectures, Physics. 1901–1921. Retrieved 2009-02-18

- Garman, E. F.; Schneider, T. R. (1997). "Macromolecular Cryocrystallography". Journal of Applied Crystallography. 30 (3): 211. doi:10.1107/S0021889897002677

- Suryanarayana, C.; Norton, M. Grant (2013-06-29). X-Ray Diffraction: A Practical Approach. Springer Science & Business Media. ISBN 9781489901484

- Jones, Nicola (2014-01-29). "Crystallography: Atomic secrets". Nature. 505 (7485): 602–603. Bibcode:-2014Natur.505..602J. doi:10.1038/505602a

- "From Atoms To Patterns". Wellcome Collection. Archived from the original on September 7, 2013. Retrieved 17 October2013

- Caspari WA (1928). "Crystallography of the Aliphatic Dicarboxylic Acids". Journal of the Chemical Society (London). ?: 3235. doi:10.1039/jr9280003235

- Harris KDM, Xu M (2009). Combined Analysis of NMR & Powder Diffraction Data. Wiley-Blackwell. ISBN 0-470-69961-2

Crystals: Structure and Symmetry

A crystal is a solid material in which its constituents are arranged in an ordered microscopic structure that extends in all directions. The study of the structure and symmetry of crystals is vital to the study of crystallography. This chapter has been carefully written to provide an easy understanding of the varied facets of crystals, such as cleavage, Bravais lattice, crystal twinning, crystallinity, etc.

Crystal

A crystal consists of matter that is formed from an ordered arrangement of atoms, molecules, or ions. The lattice that forms extends out in three-dimensions. Because there are repeated units, crystals have recognizable structures. Large crystals display flat regions (faces) and well-defined angles. Crystals with obvious flat faces are called euhedral crystals, while those lacking defined faces are called anhedral crystals. Crystals consisting of ordered arrays of atoms that aren't always periodic are called quasicrystals.

The word "crystal" comes from the Ancient Greek word krustallos, which means both "rock crystal" and "ice."

Examples of Crystals

Nonmetal Elements

The carbon atoms in diamond are held together in a covalent lattice by network covalent bonding. Iodine crystals are made of I_2 molecules held together in a crystal lattice by van der Waals forces. These forces are weak compared with covalent bonds, leading to a low melting point for iodine.

Crystalline carbon - diamond in rock.

Crystalline iodine. Image by Ben Mills.

Metal Elements

Solid metal crystals are held together by metallic bonding: i.e. a lattice of positively charged metal ions is held together by sharing delocalized conduction electrons.

Crystalline native gold. Image by Aram Dulyan.

Crystalline bismuth. Image by Heinrich Pniok.

Metalloid Elements

Crystalline silicon is held together in a covalent lattice by network covalent bonding. Tellurium crystals are formed from spiral chains of covalently bonded atoms in a hexagonal lattice.

Crystalline silicon. Image by Enricoros.

Crystalline native tellurium. Image by Christian Rewitzer.

Chemical Bonds in Crystals

The types of chemical bonds formed between atoms or groups of atoms in crystals depend on their size and electronegativity. There are four categories of crystals as grouped by their bonding:

1. Covalent Crystals - Atoms in covalent crystals are linked by covalent bonds. Pure nonmetals form covalent crystals (e.g., diamond) as do covalent compounds (e.g., zinc sulfide).

2. Molecular Crystals - Entire molecules are bonded to each other in an organized manner. A good example is a sugar crystal, which contains sucrose molecules.

3. Metallic Crystals - Metals often form metallic crystals, where some of the valence electrons are free to move throughout the lattice. Iron, for example, can form different metallic crystals.

4. Ionic Crystals - Electrostatic forces form ionic bonds. A classic example is a halite or salt crystal.

Crystal Lattices

There are seven systems of crystal structures, which are also called lattices or space lattices:

1. Cubic or Isometric - This shape includes octahedrons and dodecahedrons as well as cubes.

2. Tetragonal - These crystals form prisms and double pyramids. The structure is like a cubic crystal, except one axis is longer than the other.

3. Orthorhombic - These are rhombic prisms and dipyramids that resemble tetragons but without square cross-sections.

4. Hexagonal - Six-sided prisms with a hexagon cross section.

5. Trigonal - These crystals have a 3-fold axis.

6. Triclinic - Triclinic crystals tend not to be symmetrical.

7. Monoclinic - These crystals resemble skewed tetragonal shapes.

Lattices may have one lattice point per cell or more than one, yielding a total of 14 Bravais crystal lattice types. Bravais lattices, named for physicist and crystallographer Auguste Bravais, describe the three-dimensional array made by a set of discrete points.

A substance may form more than one crystal lattice. For example, water can form hexagonal ice (such as snowflakes), cubic ice, and rhombohedral ice. It can also form amorphous ice. Carbon can form diamond (cubic lattice) and graphite (hexagonal lattice).

Crystal Faces and Shapes

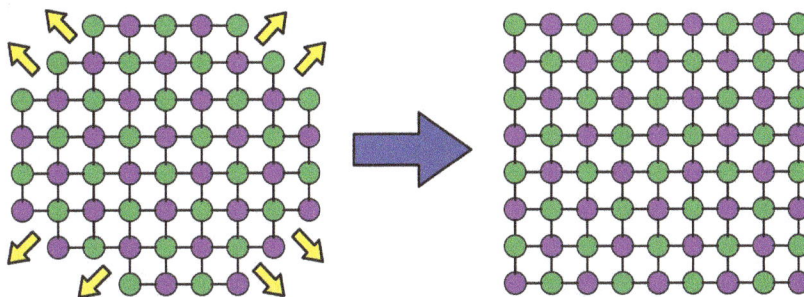

As a halite crystal is growing, new atoms can very easily attach to the parts of the surface with rough atomic-scale structure and many dangling bonds. Therefore, these parts of the crystal grow out very quickly (yellow arrows). Eventually, the whole surface consists of smooth, stable faces, where new atoms cannot as easily attach themselves.

Crystals are commonly recognized by their shape, consisting of flat faces with sharp angles. These shape characteristics are not *necessary* for a crystal—a crystal is scientifically defined by its microscopic atomic arrangement, not its macroscopic shape—but the characteristic macroscopic shape is often present and easy to see.

Euhedral crystals are those with obvious, well-formed flat faces. Anhedral crystals do not, usually because the crystal is one grain in a polycrystalline solid.

The flat faces (also called facets) of a euhedral crystal are oriented in a specific way relative to the underlying atomic arrangement of the crystal: they are planes of relatively low Miller index. This occurs because some surface orientations are more stable than others (lower surface energy). As a crystal grows, new atoms attach easily to the rougher and less stable parts of the surface, but less easily to the flat, stable surfaces. Therefore, the flat surfaces tend to grow larger and smoother, until the whole crystal surface consists of these plane surfaces.

One of the oldest techniques in the science of crystallography consists of measuring the three-dimensional orientations of the faces of a crystal, and using them to infer the underlying crystal symmetry.

A crystal's habit is its visible external shape. This is determined by the crystal structure (which restricts the possible facet orientations), the specific crystal chemistry and bonding (which may favor some facet types over others), and the conditions under which the crystal formed.

Occurrence in Nature

Ice crystals   Fossil shell with calcite crystals

Rocks

By volume and weight, the largest concentrations of crystals in the Earth are part of its solid bedrock. Crystals found in rocks typically range in size from a fraction of a millimetre to several centimetres across, although exceptionally large crystals are occasionally found. As of 1999, the world's largest known naturally occurring crystal is a crystal of beryl from Malakialina, Madagascar, 18 m (59 ft) long and 3.5 m (11 ft) in diameter, and weighing 380,000 kg (840,000 lb).

Some crystals have formed by magmatic and metamorphic processes, giving origin to large masses of crystalline rock. The vast majority of igneous rocks are formed from molten magma and the degree of crystallization depends primarily on the conditions under which they solidified. Such rocks as granite, which have cooled very slowly and under great pressures, have completely crystallized; but many kinds of lava were poured out at the surface and cooled very rapidly, and in this latter group a small amount of amorphous or glassy matter is common. Other crystalline rocks, the metamorphic rocks such as marbles, mica-schists and quartzites, are recrystallized. This means that they were at first fragmental rocks like limestone, shale and sandstone and have never been in a molten condition nor entirely in solution, but the high temperature and pressure conditions of metamorphism have acted on them by erasing their original structures and inducing recrystallization in the solid state.

Other rock crystals have formed out of precipitation from fluids, commonly water, to form druses or quartz veins. The evaporites such as halite, gypsum and some limestones have been deposited from aqueous solution, mostly owing to evaporation in arid climates.

Ice

Water-based ice in the form of snow, sea ice and glaciers is a very common manifestation of crystalline or polycrystalline matter on Earth. A single snowflake is a single crystal or a collection of crystals, while an ice cube is a polycrystal.

Organigenic Crystals

Many living organisms are able to produce crystals, for example calcite and aragonite in the case of most molluscs or hydroxylapatite in the case of vertebrates.

Polymorphism and Allotropy

The same group of atoms can often solidify in many different ways. Polymorphism is the ability of a solid to exist in more than one crystal form. For example, water ice is ordinarily found in the hexagonal form Ice I_h, but can also exist as the cubic Ice I_c, the rhombohedral ice II, and many other forms. The different polymorphs are usually called different *phases*.

In addition, the same atoms may be able to form noncrystalline phases. For example, water can also form amorphous ice, while SiO_2 can form both fused silica (an amorphous glass) and quartz (a crystal). Likewise, if a substance can form crystals, it can also form polycrystals.

For pure chemical elements, polymorphism is known as allotropy. For example, diamond and graphite are two crystalline forms of carbon, while amorphous carbon is a noncrystalline form. Polymorphs, despite having the same atoms, may have wildly different properties. For example, diamond is among the hardest substances known, while graphite is so soft that it is used as a lubricant.

Polyamorphism is a similar phenomenon where the same atoms can exist in more than one amorphous solid form.

Crystallization

Crystallization is the process of forming a crystalline structure from a fluid or from materials dissolved in a fluid.

Vertical cooling crystallizer in a beet sugar factory.

Crystallization is a complex and extensively-studied field, because depending on the conditions, a single fluid can solidify into many different possible forms. It can form a single crystal, perhaps with various possible phases, stoichiometries, impurities, defects, and habits. Or, it can form a polycrystal, with various possibilities for the size, arrangement, orientation, and phase of its grains. The final form of the solid is determined by the conditions under which the fluid is being solidified, such as the chemistry of the fluid, the ambient pressure, the temperature, and the speed with which all these parameters are changing.

Specific industrial techniques to produce large single crystals (called *boules*) include the Czochralski process and the Bridgman technique. Other less exotic methods of crystallization may be used, depending on the physical properties of the substance, including hydrothermal synthesis, sublimation, or simply solvent-based crystallization.

Large single crystals can be created by geological processes. For example, selenite crystals in excess of 10 meters are found in the Cave of the Crystals in Naica, Mexico.

Crystals can also be formed by biological processes. Conversely, some organisms have special techniques to *prevent* crystallization from occurring, such as antifreeze proteins.

Defects, Impurities, and Twinning

An *ideal* crystal has every atom in a perfect, exactly repeating pattern. However, in reality, most crystalline materials have a variety of crystallographic defects, places where the crystal's pattern is interrupted. The types and structures of these defects may have a profound effect on the properties of the materials.

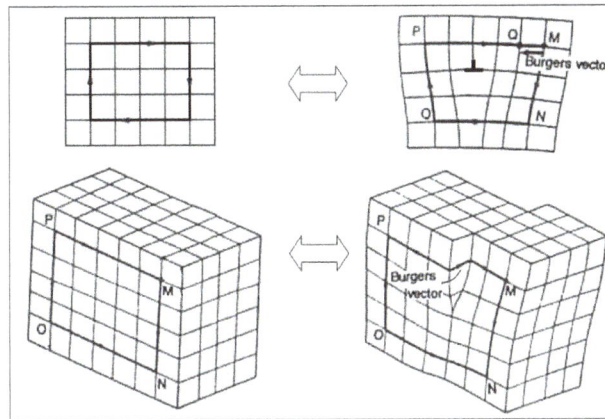

Two types of crystallographic defects. Top right: edge dislocation. Bottom right: screw dislocation.

A few examples of crystallographic defects include vacancy defects (an empty space where an atom should fit), interstitial defects (an extra atom squeezed in where it does not fit), and dislocations. Dislocations are especially important in materials science, because they help determine the mechanical strength of materials.

Another common type of crystallographic defect is an impurity, meaning that the "wrong" type of atom is present in a crystal. For example, a perfect crystal of diamond would only contain carbon atoms, but a real crystal might perhaps contain a few boron atoms as well. These boron impurities change the diamond's color to slightly blue. Likewise, the only difference between ruby and sapphire is the type of impurities present in a corundum crystal.

Twinned pyrite crystal group.

In semiconductors, a special type of impurity, called a dopant, drastically changes the crystal's electrical properties. Semiconductor devices, such as transistors, are made possible largely by putting different semiconductor dopants into different places, in specific patterns.

Twinning is a phenomenon somewhere between a crystallographic defect and a grain boundary. Like a grain boundary, a twin boundary has different crystal orientations on its two sides. But unlike a grain boundary, the orientations are not random, but related in a specific, mirror-image way.

Mosaicity is a spread of crystal plane orientations. A mosaic crystal is supposed to consist of smaller crystalline units that are somewhat misaligned with respect to each other.

Crystal Structure

In mineralogy and crystallography, crystal structure is a unique arrangement of atoms or molecules in a crystalline liquid or solid. A crystal structure is composed of a pattern, a set of atoms arranged in a particular way, and a lattice exhibiting long-range order and symmetry. Patterns are located upon the points of a lattice, which is an array of points repeating periodically in three dimensions. The points can be thought of as forming identical tiny boxes, called unit cells, that fill the space of the lattice. The lengths of the edges of a unit cell and the angles between them are called the lattice parameters. The symmetry properties of the crystal are embodied in its space group.

A crystal's structure and symmetry play a role in determining many of its physical properties, such as cleavage, electronic band structure, and optical transparency.

Cleavage

Cleavage, tendency of a crystalline substance to split into fragments bounded by plane surfaces. Although cleavage surfaces are seldom as flat as crystal faces, the angles between them are highly characteristic and valuable in identifying a crystalline material.

Cleavage occurs on planes where the bonding forces are weakest. A crystal may be cleaved with equal ease in any direction that is parallel to crystallographically identical faces; for example, galena cleaves parallel to all faces of a cube. Cleavage is described by its direction (as cubic, prismatic, basal) and by the ease with which it is produced. A perfect cleavage produces smooth, lustrous surfaces with great ease. Other degrees include distinct, imperfect, and difficult.

Cleavage is often measured by three factors:

1) Quality of Cleavage

2) Number of Sides Exhibiting Cleavage

3) Cleavage Habit

Quality of Cleavage

Quality of cleavage can be categorized into five qualities:

- Perfect

- Good

- Poor

- Indiscernible (Indistinct)

- None

Minerals with perfect cleavage will cleave without leaving any rough surfaces; a full, smooth plane is formed where the crystal broke. Minerals with good cleavage also leave smooth surfaces, but often leave over minor residual rough surfaces. On minerals with poor cleavage, the smooth crystal edge is not very visible, since the rough surface is dominant. If a mineral exhibits cleavage, but it so poor that it is hardly noticeable, it has "indiscernible" cleavage. Minerals with no cleavage never exhibit any cleavage, thus broken surfaces are fractured and rough.

Categorization of cleavage qualities is not scientifically affirmed. The above categorization is used by most mineral references, but some guides categorize cleavage in three or four groups, and may give them different names, such as "excellent" and "distinct".

Number of Sides Exhibiting Cleavage

Many minerals exhibit cleavage only on one side, and some may exhibit different quality cleavage on different crystal sides. The following criteria may be expected when analyzing the cleavage of any particular mineral:

- One Direction
- Two Directions
- Three Directions
- All Directions

These identify how many "directions", or planes, the crystal is exhibiting the cleavage on. Each direction signifies the two opposite sides of a three-dimensional figure, (since opposite sides will always exhibit the same cleavage properties). If a mineral has cleavage in three directions, then every side of the mineral has cleavage (i.e. length, width, and height). If a mineral occurs in modified crystals with more than six sides (i.e. an octahedron) and exhibits cleavage on all the sides, than it has cleavage in "all directions".

Combining the cleavage level together with the number of sides will measure the cleavage of a mineral. For example, if a mineral has Good Cleavage, Two Directions, this means that it has good cleavage on four out of six sides (while the other two sides exhibit no cleavage). If a mineral has Perfect Cleavage, One Direction; Poor Cleavage, Two Directions, this means that the mineral has perfect cleavage on two sides, and poor cleavage on the other four.

In this guide, cleavage quality is measured in numbers, then the amount of sides, separated by a comma. 1 is perfect cleavage, 2 is good cleavage, and 3 is poor cleavage. If the cleavage of a mineral is written as 1,2 the mineral has perfect cleavage in two directions. If all sides of mineral have the same cleavage, and the mineral often occurs in modified crystals with more than six sides, than All Sides is written instead of a number. If a mineral exhibits different cleavage on different crystal planes, there will be two cleavage indicators separated by a semi-colon (;). For example, if the cleavage of a mineral is written as 1,2;3,1, than it has perfect cleavage in two directions, and poor cleavage in one other direction. If a mineral exhibits indistinct or no cleavage, Indiscernible or None is written in the cleavage field.

Cleavage Habit

Different habits of cleavage exist on different minerals, depending on their mode of crystallization. These forms of cleavage are:

Basal cleavage:

> Cleavage exhibited on a horizontal plane of the mineral by way of its base. Minerals with basal cleavage can sometimes be "peeled".

> An example of basal cleavage are the mica minerals.

Cubic cleavage:

> Cleavage exhibited on minerals of the isometric crystal system that are crystallized as cubes. In this method of cleavage, small cubes evenly break off of an existing cube.

> An example is Galena.

Octahedral cleavage:

> Cleavage exhibited on minerals of the isometric crystal system that are crystallized as octahedrons. In this method of cleavage, flat, triangular "wedges" peel off of an existing octahedron.

> An example is Fluorite.

Prismatic cleavage:

> Cleavage exhibited on some prismatic minerals in which a crystal cleaves as thin, vertical, prismatic crystals off of the original prism.

> An example is Aegirine.

Pinicoidal cleavage:

> Cleavage exhibited on some prismatic and tabular minerals in which a crystal cleaves on the pinacoidal plane, which is the third dimension aside from the basal and prismatic sides.

> An example is Barite.

Rhombohedral cleavage:

> Cleavage exhibited on minerals crystallizing in the hexagonal crystal system as rhombohedrons, in which small rhombohedrons break off of the existing rhombohedron.

> An example is Calcite.

Parting:

> Parting is characteristically similar to cleavage. It is easily confused with cleavage, and it may be present on minerals that do not exhibit any cleavage. There are two causes of parting:

> 1) Two separate pressures pushed toward the center of a crystal after its formation, causing the crystal interior to evenly dislodge on a flat, smooth plane.

> 2) Twinned crystals that separated from one another, leaving a flat, smooth plane.

> With enough perception, a distinction can be made between parting and cleavage. If fracture marks are present on a crystal in addition to a cleaved plane, the "cleaved" surface is

usually the result of parting, not cleavage. An outline of a crystal etched in a mineral is also the result of parting, in the form of twinned crystals that separated.

In general, one need not worry about confusing parting with cleavage. Parting is uncommon, and it can usually be determined by the distinguishing characteristics mentioned above.

Fracture:

Fracture is the characteristic mark left when a mineral chips or breaks. Cleavage and fracture differ in that cleavage is the break of a crystal face where a new face (resulting in a smooth plane) is formed, whereas fracture is the "chipping" shape of a mineral. All minerals exhibit a fracture, even those that exhibit cleavage. If a mineral with cleavage is chipped a certain way, it will fracture rather than cleave.

There are several terms to describe the various mineral fractures:

Conchoidal - Fracture resembling a semicircular shell, with a smooth, curved surface. An example of conchoidal fracture can be seen in broken glass. (This fracture is also known as "shelly" in some reference guides.)

Uneven - Fracture that leaves a rough or irregular surface.

Hackly - Fracture that resembles broken metal, with rough, jagged, points. True metals exhibit this fracture. (This fracture is also known as "jagged".)

Splintery - Fracture that forms elongated splinters. All fibrous minerals fall into this ctegory.

Earthy or crumbly - Fracture of minerals that crumble when broken.

Even or smooth - Fracture that forms a smooth surface.

Subconchoidal - Fracture that falls somewhere between conchoidal and even; smooth with irregular rounded corners.

Some references may describe additional fractures not mentioned above, but those terms are either synonymous or simply used as a verbal depiction of the authors inference.

Almost all minerals have a characteristic fracture. Some minerals of the same species may exhibit a different fracture, but this is rare.

Crystallinity

Crystallinity defines the degree of long-range order in a material, and strongly affects its properties. The more crystalline a polymer, the more regularly aligned its chains. Increasing the degree of crystallinity increases hardness and density. This is illustrated in poly(ethene).

HDPE (high density poly(ethene)) is composed of linear chains with little branching. Molecules pack closely together, leading to a high degree of order. This makes it stiff and dense, and it is used for milk bottles and drainpipes.

The numerous short branches in LDPE (low density poly(ethene)) interfere with the close packing of molecules, so they cannot form an ordered structure. The lower density and stiffness make it suitable for use in films such as plastic carrier bags and food wrapping.

Often, polymers are semi-crystalline, existing somewhere on a scale between amorphous and crystalline. This usually consists of small crystalline regions (crystallites) surrounded by regions of amorphous polymer.

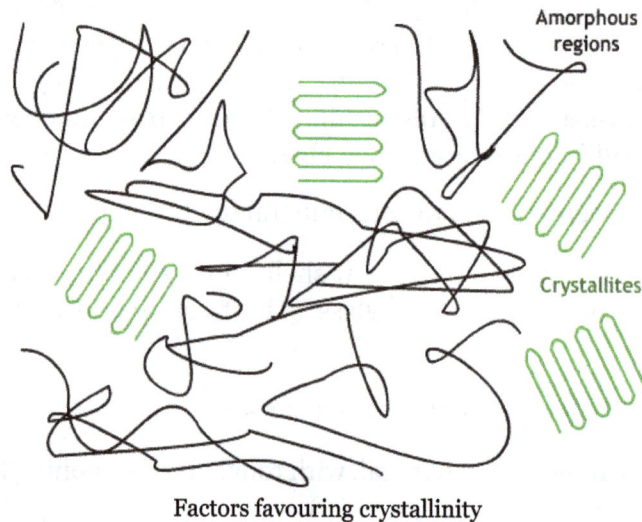

Factors favouring crystallinity

In general, factors causing polymers to be more ordered and regular tend to lead to a higher degree of crystallinity.

- Fewer short branches – allowing molecules to pack closely together.

- Higher degree of stereoregularity - syndiotactic and isotactic polymers are more ordered than atactic polymers.

- More regular copolymer configuration – having the same effect as stereoregularity.

This topic is covered in the Crystallinity in Polymers TLP.

Crystallinity generally is described in terms of the four categories shown in the Table.

Crystallinity categories of igneous rocks	
Crystallinity	Rock Term
Entirely Crystalline	Holocrystalline
Crystalline Material and subordinate glass	Hemicrystalline or hypocrystalline
Glass and subordinate crystalline material	Hemihyaline or hypohyaline
Entirely Glassy	Holohyaline or hyaline

Those holocrystalline rocks in which mineral grains can be recognized with the unaided eye are called phanerites, and their texture is called phaneritic. Those with mineral grains so small that their outlines cannot be resolved without the aid of a hand lens or microscope are termed aphanites, and their texture is termed aphanitic. Aphanitic rocks are further described as either microcrystalline or cryptocrystalline, according to whether or not their individual constituents can be resolved under the microscope. The subaphanitic, or hyaline, rocks are referred to as glassy, or vitric, in terms of granularity.

Aphanitic and glassy textures represent relatively rapid cooling of magma and, hence, are found mainly among the volcanic rocks. Slower cooling, either beneath Earth's surface or within very thick masses of lava, promotes the formation of crystals and, under favourable circumstances of magma composition and other factors, their growth to relatively large sizes. The resulting phaneritic rocks are so widespread and so varied that it is convenient to specify their grain size as shown in the Table.

Categories of rock grain size		
terms in common use	general grain size	
	igneous rocks in general	pegmatites
fine-grained	<1 mm	<1 in.
medium-grained	1–5 mm	1 in.–4 in.
coarse-grained	5 mm–2 cm	4 in.–12 in.
very coarse-grained	>2 cm	>12 in.

Bravais Lattice

The Bravais lattice are the distinct lattice types which when repeated can fill the whole space. The lattice can therefore be generated by three unit vectors, a_1, a_2 and a_3 and a set of integers k, l and m so that each lattice point, identified by a vector r, can be obtained from:

$$r = k\,a_1 + l\,a_2 + m\,a_3$$

In two dimensions there are five distinct Bravais lattices, while in three dimensions there are fourteen. These fourteen lattices are further classified as shown in the table below where a_1, a_2 and a_3 are the magnitudes of the unit vectors and a, b and g are the angles between the unit vectors.

Name	Number of Bravais lattices	Conditions
Triclinic	1	$a_1 \neq a_2 \neq a_3$ $\alpha \neq \beta \neq \gamma$
Monoclinic	2	$a_1 \neq a_2 \neq a_3$ $\alpha = \beta = 90° \neq \gamma$
Orthorhombic	4	$a_1 \neq a_2 \neq a_3$ $\alpha = \beta = \gamma = 90°$
Tetragonal	2	$a_1 = a_2 \neq a_3$ $\alpha = \beta = \gamma = 90°$
Cubic	3	$a_1 = a_2 = a_3$ $\alpha = \beta = \gamma = 90°$
Trigonal	1	$a_1 = a_2 = a_3$ $\alpha = \beta = \gamma < 120° \neq 90°$
Hexagonal	1	$a_1 = a_2 \neq a_3$ $\alpha = \beta = 90°$ $\gamma = 120°$

The 14 Bravais Lattices

So one classifies different lattices according to the shape of the parallelepiped spanned by its primitive translation vectors.

However, this is not yet the best solution for a classification with respect to symmetry. Consider for example the unit cells (a) and (b) presented before: While cell (a) is the actual unit cellspanned by the primitive translation vectors, it does not show the symmetry of the latticeproperly whereas cell (b) clearly shows the two axes of rotation.

So sometimes it makes sense not to use a primitive unit cell but one which fits better to the symmetry of the problem. This idea leads to the 14 Bravais Lattices which are depicted below ordered by the crystal systems:

Cubic

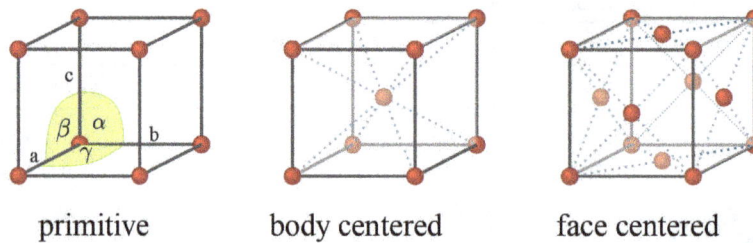

primitive   body centered   face centered

There are three Bravais lattices with a cubic symmetry. One distinguishes the simple/primitive cubic (sc), the body centered cubic (bcc) and the face centered cubic (fcc) lattice.

Tetragonal

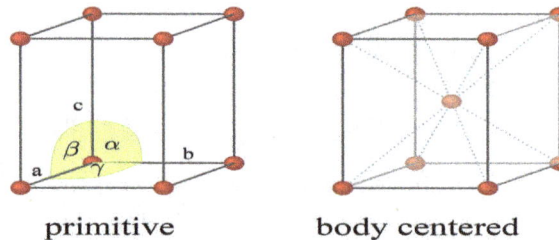

primitive   body centered

There are two tetragonal Bravais lattices with $a=b\neq c$ and $\alpha=\beta=\gamma=90°$. One is primitive and the other body centered.

Orthorhombic

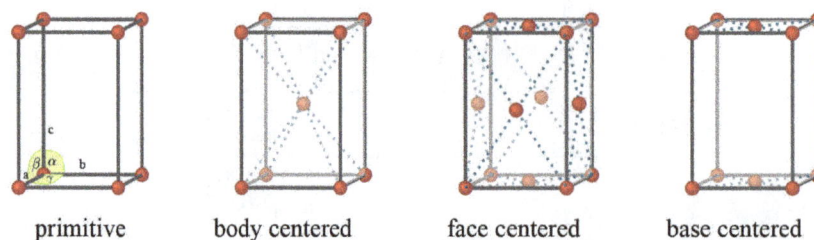

primitive   body centered   face centered   base centered

There are four orthorhombic Bravais lattices with a≠b≠c and α=β=γ=90°: Primitive, body centered, face centered and base centered.

Hexagonal3

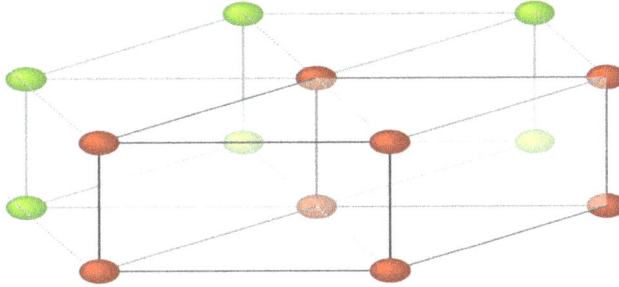

When two sides are of equal length with an enclosed angle of 120° the crystal has a hexagonal structure and thus a 6-fold rotary axis.

Monoclinic

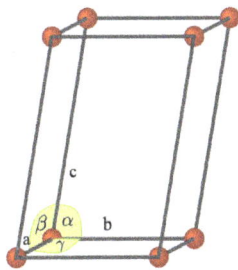 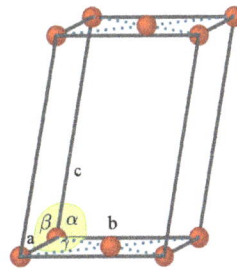

primitive   base centered

As in the orthohombric structure, all edges are of unequal length. However, one of the three angles is ≠90°.

Trigonal and Triclinic

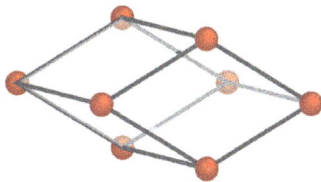 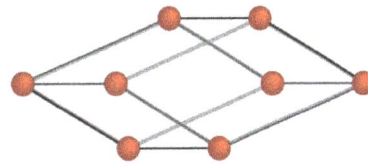

trigonal   triclinic

The trigonal (or rhombohedral) lattice has three edges of equal length and three equal angles (≠90°). In the triclinic lattice, all edges and angles are unequal.

In 2 Dimensions

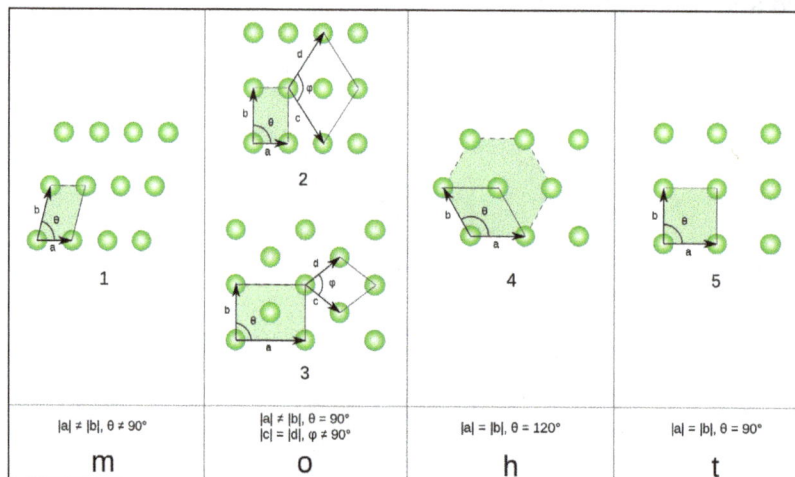

$|a| \neq |b|, \theta \neq 90°$	$|a| \neq |b|, \theta = 90°$ $|c| = |d|, \varphi \neq 90°$	$|a| = |b|, \theta = 120°$	$|a| = |b|, \theta = 90°$
m	o	h	t

1 – oblique (monoclinic), 2 – rectangular (orthorhombic), 3 – centered rectangular (orthorhombic), 4 – hexagonal, and 5 – square (tetragonal).

In two-dimensional space, there are 5 Bravais lattices, grouped into four crystal families.

Crystal family	Schönflies	5 Bravais lattices	
		Primitive	Centered
Monoclinic	C_2	Oblique	
Orthorhombic	D_2	Rectangular	Centered rectangular
Hexagonal	D_6	Hexagonal	
Tetragonal	D_4	Square	

The unit cells are specified according to the relative lengths of the cell edges (a and b) and the angle between them (θ). The area of the unit cell can be calculated by evaluating the norm $||a \times b||$, where a and b are the lattice vectors. The properties of the crystal families are given below:

Crystal family	Area	Axial distances (edge lengths)	Axial angle
Monoclinic		$a \neq b$	$\theta \neq 90°$
Orthorhombic		$a \neq b$	$\theta = 90°$
Hexagonal		$a = b$	$\theta = 120°$
Tetragonal		$a = b$	$\theta = 90°$

In 3 Dimensions

In three-dimensional space, there are 14 Bravais lattices. These are obtained by combining one of the seven lattice systems with one of the centering types. The centering types identify the locations of the lattice points in the unit cell as follows:

- Primitive (P): lattice points on the cell corners only (sometimes called simple).

- Base-centered (A, B, or C): lattice points on the cell corners with one additional point at the center of each face of one pair of parallel faces of the cell (sometimes called end-centered).

- Body-centered (I): lattice points on the cell corners, with one additional point at the center of the cell.

- Face-centered (F): lattice points on the cell corners, with one additional point at the center of each of the faces of the cell.

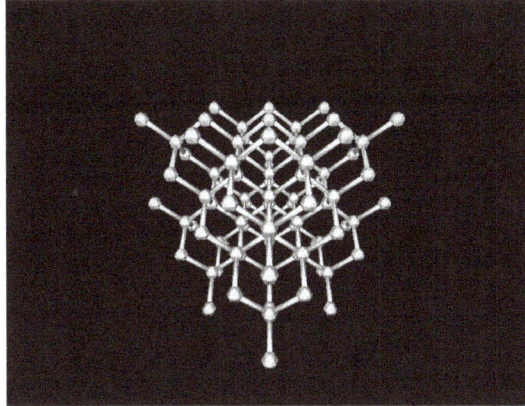

2×2×2 unit cells of a diamond cubic lattice

Not all combinations of lattice systems and centering types are needed to describe all of the possible lattices, as it can be shown that several of these are in fact equivalent to each other. For example, the monoclinic I lattice can be described by a monoclinic C lattice by different choice of crystal axes. Similarly, all A- or B-centred lattices can be described either by a C- or P-centering. This reduces the number of combinations to 14 conventional Bravais lattices, shown in the table below.

Crystal Family	Lattice System	Schönflies	14 Bravais Lattices			
			Primitive (P)	Base-centered (C)	Body-centered (I)	Face-centered (F)
Triclinic		C_i	γ β c a α b			
Monoclinic		C_{2h}	$\beta \neq 90°$ $a \neq c$ β c a b	$\beta \neq 90°$ $a \neq c$ β c a b		
Orthorhombic		D_{2h}	$a \neq b \neq c$ c a b	$a \neq b \neq c$ c a b	$a \neq b \neq c$ c a b	$a \neq b \neq c$ c a b

	Lattice system	Symmetry			
Tetragonal		D_{4h}	$a \neq c$		$a \neq c$
Hexagonal	Rhombo-hedral	D_{3d}	$\alpha \neq 90°$		
	Hexagonal	D_{6h}	$\gamma = 120°$		
Cubic		O_h			

The unit cells are specified according to the relative lengths of the cell edges (a, b, c) and the angles between them (α, β, γ). The volume of the unit cell can be calculated by evaluating the triple product a · (b × c), where a, b, and c are the lattice vectors. The properties of the lattice systems are given below:

Crystal family	Lattice system	Volume	Axial distances (edge lengths)	Axial angles	Corresponding examples
Triclinic		$abc\sqrt{1 - \cos^2\alpha - \cos^2\beta - \cos^2\gamma + 2\cos\alpha\cos\beta\cos\gamma}$	(All remaining cases)		$K_2Cr_2O_7$, $CuSO_4 \cdot 5H_2O$, H_3BO_3
Monoclinic		$abc\sin\beta$	$a \neq c$	$\alpha = \gamma = 90°, \beta \neq 90°$	Monoclinic sulphur, $Na_2SO_4 \cdot 10H_2O$, $PbCrO_3$
Orthorhombic		abc	$a \neq b \neq c$	$\alpha = \beta = \gamma = 90°$	Rhombic sulphur, KNO_3, $BaSO_4$
Tetragonal		a^2c	$a = b \neq c$	$\alpha = \beta = \gamma = 90°$	White tin, SnO_2, TiO_2, $CaSO_4$

Hexagonal	Rhombohedral	$a^3\sqrt{1-3\cos^2\alpha+2\cos^3\alpha}$	$a=b=c$	$\alpha=\beta=\gamma\neq 90°$	Calcite (CaCO$_3$), cinnabar (HgS)
	Hexagonal	$\dfrac{\sqrt{3}}{2}a^2c$	$a=b$	$\alpha=\beta=90°$, $\gamma=120°$	Graphite, ZnO, CdS
Cubic		a^3	$a=b=c$	$\alpha=\beta=\gamma=90°$	NaCl, zinc blende, copper metal, KCl, Diamond, Silver

In 4 Dimensions

In four dimensions, there are 64 Bravais lattices. Of these, 23 are primitive and 41 are centered. Ten Bravais lattices split into enantiomorphic pairs.

Crystal Twinning

Twinning, in crystallography, regular intergrowth of two or more crystal grains so that each grain is a reflected image of its neighbour or is rotated with respect to it. Other grains added to the twin form crystals that often appear symmetrically joined, sometimes in a starlike or cross-like shape.

Twinning often occurs from the beginning of crystal growth. The individuals that comprise a twin have atomic structures with different orientations, but they must have certain common planes or directions. They must fit simply and must be derived from each other by a simple movement.

There are several kinds of twin crystals. Penetration twins are complete crystals that pass through one another and often share the centre of their axial systems.

Some geometric relations concerning crystal twinning can be set down. Twinning results in re-flected images along a common twinning plane, repetitions rotated about a common twinning axis, or both. Such twinning planes and axes have simple relations to the crystallographic axes of the crystal and are governed by some fundamental laws; e.g., because the resulting twin would be identical to the original crystal, no plane of symmetry in the simple crystal may become a twinning plane, and no axis of 2-, 4-, or 6-fold symmetry may become a twinning axis; also, twinned crystals in classes with a centre of symmetry will have a twinning axis perpendicular to a twinning plane, but, lacking a centre of symmetry, a twinning axis or plane may occur independently.

Twin Laws

Twin laws are expressed as either form symbols to define twin planes (i.e. {hkl}) or zone symbols to define the direction of twin axes (i.e. [hkl]).

The surface along which the lattice points are shared in twinned crystals is called a *composition surface*.

If the twin law can be defined by a simple planar composition surface, the twin plane is alwaysparallel to a possible crystal face and never parallel to an existing plane of symmetry (remember that twinning adds symmetry).

If the twin law is a rotation axis, the composition surface will be irregular, the twin axis will be perpendicular to a lattice plane, but will never be an even-fold rotation axis of the existing symmetry. For example twinning cannot occur on a new 2 fold axis that is parallel to an existing 4-fold axis.

Types of Twinning

Another way of defining twinning breaks twins into two separate types.

1. *Contact Twins* - have a planar composition surface separating 2 individual crystals. These are usually defined by a twin law that expresses a twin plane (i.e. an added mirror plane). An example shown here is a crystal of orthoclase twinned on the Braveno Law, with {021} as the twin plane.	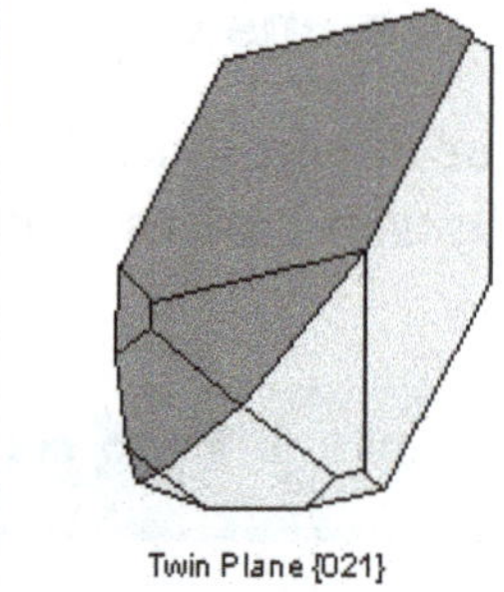 Twin Plane {021}
2. *Penetration Twins* - have an irregular composition surface separating 2 individual crystals. These are defined by a twin center or twin axis. Shown here is a twinned crystal of orthoclase twinned on the Carlsbad Law with [001] as the twin axis.	 [001]
Contact twins can also occur as repeated or multiple twins. • If the compositions surfaces are parallel to one another, they are called *polysynthetic twins*. Plagioclase commonly shows this type of twinning, called the Albite Twin Law, with {010} as the twin plane. Such twinning is one of the most diagnostic features of plagioclase.	 (010)
• If the composition surfaces are not parallel to one another, they are called *cyclical twins*. Shown here is the cyclical twin that occurs in chrysoberyl along a {031} plane.	 {031}

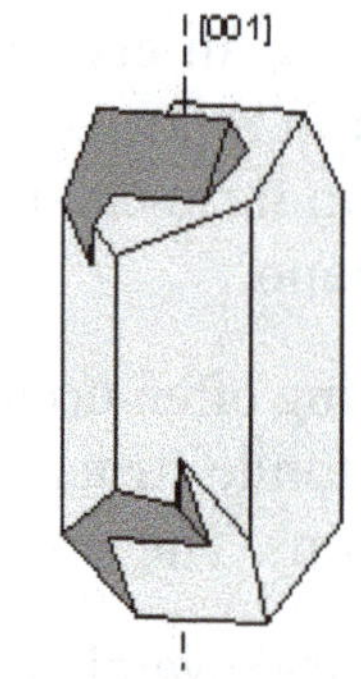

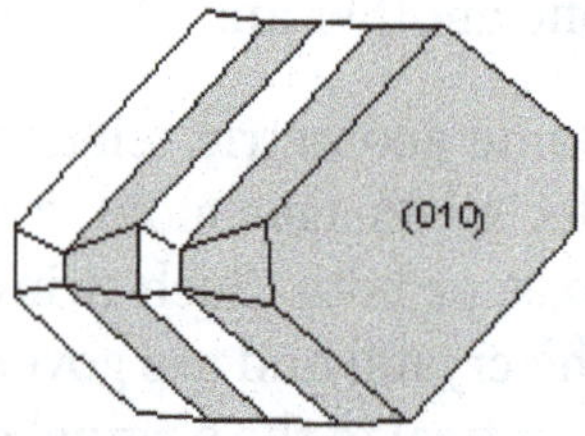

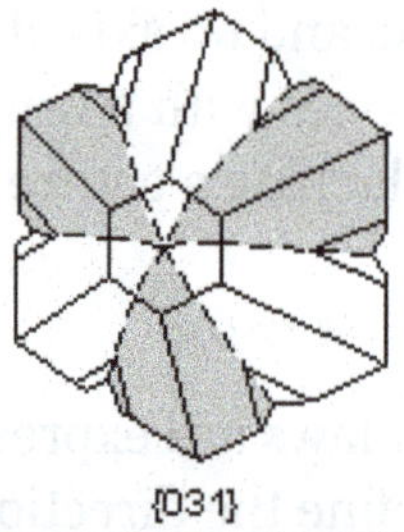

Modes of Formation

There are three modes of formation of twinned crystals. *Growth twins* are the result of an interruption or change in the lattice during formation or growth due to a possible deformation from a larger substituting ion. *Annealing* or *transformation twins* are the result of a change in crystal system during cooling as one *form* becomes unstable and the crystal structure must re-organize or *transform* into another more stable form. *Deformation* or *gliding twins* are the result of stress on the crystal after the crystal has formed. If an *FCC* metal like aluminium experiences extreme stresses, it will experience twinning, as seen in the case of explosions. Deformation twinning is a common result of regional metamorphism. Crystal twinning is also used as an indicator of force direction in mountain building processes in orogeny research.

Crystals that grow adjacent to each other may be aligned to resemble twinning. This *parallel growth* simply reduces system energy and is not twinning.

Twin boundaries can be constructed from both twist and tilt boundaries.

Mechanisms of Formation

Twinning can occur cooperative displacement of atoms along the face of the twin boundary. This displacement of a large quantity of atoms simultaneously requires significant energy to perform. Therefore, the theoretical stress required to form a twin is quite high. It is believed that twinning is associated with dislocation motion on a coordinated scale, in contrast to slip, which is caused by independent glide at several locations in the crystal.

Twinning and slip are competitive mechanisms for crystal deformation. Each mechanism is dominant in certain crystal systems and under certain conditions. In FCC metals, slip is almost always dominant because the stress required is far less than twinning stress.

Compared to slip, twinning produces a deformation pattern that is more heterogeneous in nature. This deformation produces a local gradient across the material and near intersections between twins and grain boundaries. The deformation gradient can lead to fracture along the boundaries, particularly in BCC transition metals at low temperatures.

Deposition of Twins

The conditions of crystal formation in solution have an effect on the type and density of dislocations in the crystal. It frequently happens that the crystal is oriented so that there will a more rapid deposition of material on one part than on another; for instance, if the crystal be attached to some other solid it cannot grow in that direction. If the crystal is freely suspended in the solution and material for growth is supplied at the same rate on all sides does an equably developed form result.

Twin Boundaries

Twin boundaries occur when two crystals of the same type intergrow so that only a slight misorientation exists between them. It is a highly symmetrical interface, often with one crystal the mirror image of the other; also, atoms are shared by the two crystals at regular intervals. This is also a much lower-energy interface than the grain boundaries that form when crystals of arbitrary orientation grow together. Twin boundaries may also display a higher degree of symmetry than the single crystal. These twins are called *mimetic* or *pseudo-symmetric* twins.

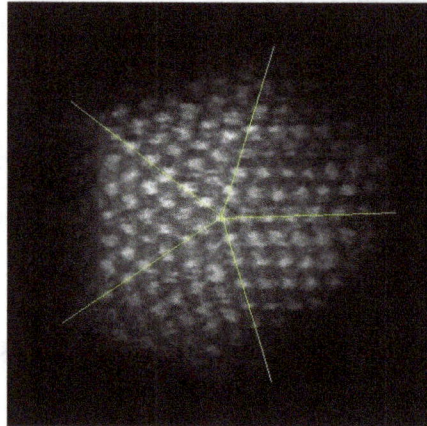

Fivefold twinning in a gold nanoparticle (electron microscope image).

Twin boundaries are partly responsible for shock hardening and for many of the changes that occur in cold work of metals with limited slip systems or at very low temperatures. They also occur due to martensitic transformations: the motion of twin boundaries is responsible for the pseudoelastic and shape-memory behavior of nitinol, and their presence is partly responsible for the hardness due to quenching of steel. In certain types of high strength steels, very fine deformation twins act as primary obstacles against dislocation motion. These steels are referred to as 'TWIP' steels, where TWIP stands for *Twinning Induced Plasticity*.

Appearance in Different Structures

Of the three common crystalline structures BCC, FCC, and HCP, the HCP structure is the most likely to form deformation twins when strained, because they rarely have a sufficient number of slip systems for an arbitrary shape change. High strain rates, low stacking-fault energy and low temperatures facilitate deformation twinning.

Centrosymmetry

The term centrosymmetric refers to a space group which contains an inversion center as one of its symmetry elements *i.e.* for every point (x, y, z) in the unit cell, there is an indistinguishable point (-x, -y, -z).

Crystal class	Centro symmetric Point groups		Noncentrosymmetric Point groups				
			Polar		Non-polar		
Cubic	m3	m3m	none	432	$\bar{3}$ m		23
Tetragonal	4 or m	4 or mmm	4	4mm	$\bar{4}$	$\bar{4}$ 2 m	22
Orthorhombic		mmm	mm2		222		

Hexagonal	6 or m	6 or mmm	6	6mm	$\bar{6}$	$\bar{6}2m$	622
Trigonal	$\bar{3}$	$\bar{3}$ m	3	3m			32
Monoclinic	2 or m		2	m			none
Triclinic	$\bar{1}$		1				none
Total Number	11 groups		10 groups				11 groups

Out of these 21 point groups, except group 432, crystals containing all other point groups exhibit piezoelectric effect *i.e.* upon application of an electric field, they exhibit strain or upon application of an external stress, charges develop on the faces of crystal resulting in an induced electric field.

Out of these 20 non-centrosymmetric point groups, 10 belong to polar crystals *i.e.* crystals which possess a unique polar axis, an axis showing different properties at the two ends.

These crystals can be spontaneously polarized and polarization can be compensated through external or internal conductivity or twinning or domain formation.

Spontaneous polarization depends upon the temperature. Consequently, if a change in temperature is imposed, an electric charge is developed on the faces of the crystal perpendicular to the polar axis. This is called pyroelectric effect. All 10 classes of polar crystals are pyroelectric.

In some of these polar non-centrosymmetric crystals, the polarization along the polar axis can be reversed by reversing the polarity of electric field. Such crystals are called ferroelectric i.e. these are spontaneously polarized materials with reversible polarization.

So by default, all ferroelectric materials are simultaneously pyroelectric and piezoelectric. Similarly, all pyroelectric materials are by default piezoelectric but not all of them are ferroelectric.

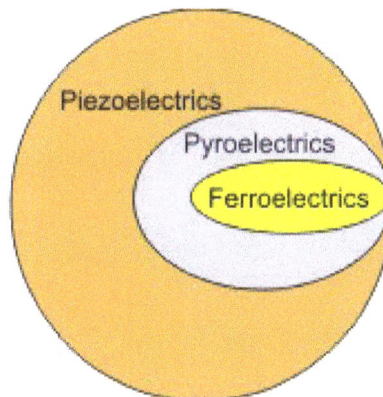

Classificatin of piezo-, pyro- and ferro-electrics

References

- Libbrecht, Kenneth; Wing, Rachel (2015-09-01). The Snowflake: Winter's Frozen Artistry. Voyageur Press. ISBN 9781627887335

- International Union of Crystallography (1992). "Report of the Executive Committee for 1991". Acta Crystallogr. A. 48 (6): 922. doi:10.1107/S0108767392008328. PMC 1826680

- "The International Annealed Copper Standard". Nondestructive Testing Resource Center. The Collaboration for NDT Education, Iowa State University. n.d. Retrieved November 14, 2016

- Brown, Harold; Bülow, Rolf; Neubüser, Joachim; Wondratschek, Hans; Zassenhaus, Hans (1978), Crystallographic groups of four-dimensional space, New York: Wiley-Interscience [John Wiley & Sons], ISBN 978-0-471-03095-9, MR 0484179

- Aroyo, Mois I.; Müller, Ulrich; Wondratschek, Hans (2006). "Historical Introduction". International Tables for Crystallography. Springer. A1(1.1): 2–5. doi:10.1107/97809553602060000537. Archived from the original on 2013-07-04. Retrieved 2008-04-21

Crystal Systems and its Types

In the field of crystallography, the classes of crystals that are classified with respect to the inclination of the crystal axes or respective symmetries of the crystal are called crystal systems. There are seven such categories of crystals, namely cubic, hexagonal, trigonal, tetragonal, orthorhombic, monoclinic and triclinic crystal systems, which have been discussed in detail in this chapter.

Crystal System

Crystal system is a method of classifying crystalline substances on the basis of their unit cell. There are seven unique crystal systems. The simplest and most symmetric, the cubic (or isometric) system, has the symmetry of a cube. The other six systems, in order of decreasing symmetry, are hexagonal, tetragonal, trigonal (also known as rhombohedral), orthorhombic, monoclinic and triclinic.

Bravais lattice is a set of points constructed by translating a single point in discrete steps by a set of basis vectors. In 1848, the French physicist and crystallographer Auguste Bravais (1811-1863) established that in three-dimensional space only fourteen different lattices may be constructed. All crystalline materials recognised till now fit in one of these arrangements.

Cubic Crystal System

Crystallographic Axes

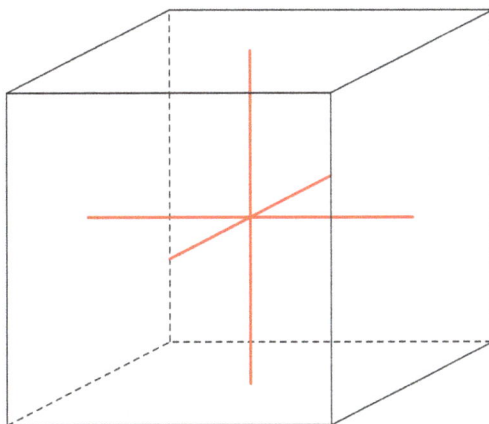

Cubic (or isometric) crystal system is also known as the isometric system. Cubic crystal system characterizes itself by its three equivalent crystallographic axes perpendicular to each other.

$a = b = c$

$\alpha = \beta = \gamma = 90°$

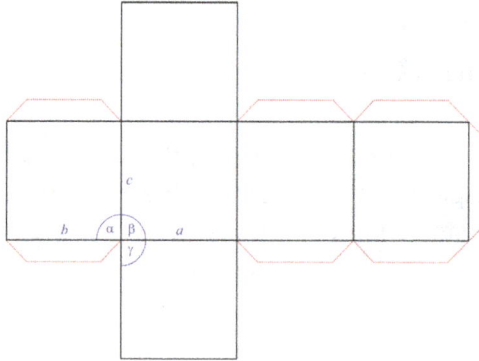

Volume of unit cell
$V = a^3$
Bravais lattices

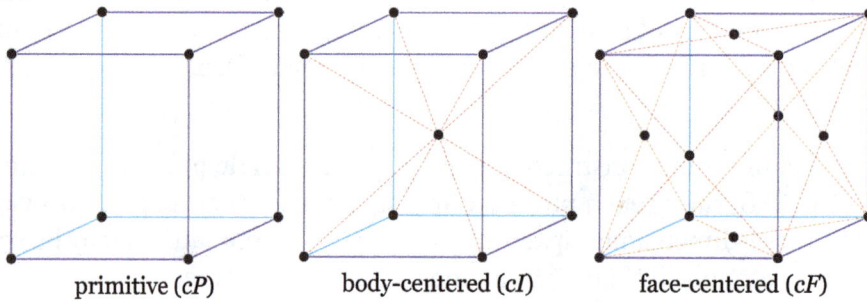

primitive (cP)	body-centered (cI)	face-centered (cF)

Hexagonal Crystal System

Crystallographic Axes

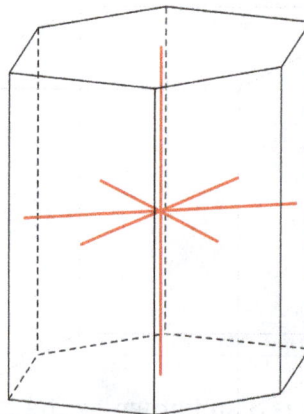

Hexagonal crystal system is based on four crystallographic axes. Three of the axes (denoted by a_1, a_2, and a_3) are of the same length and lie in the hexagonal (basal) plane at 120° to one another (between the positive ends). A fourth axis (c), longer or shorter than other three, is perpendicular to this plane. Therefore, it is sufficient to give the a and c lattice parameters for the descriptioion of the hexagonal lattice.

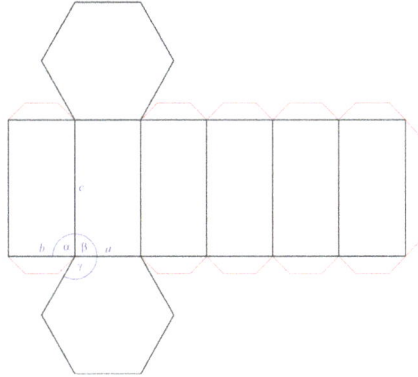

Hexagonal Unit Cell

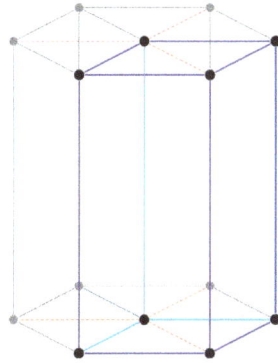

The unit cell parameters are:

$a = b \neq c$

$\alpha = \beta = 90° \gamma = 120°$

The volume of the hexagonal unit cell is given by

$V = a^2c \sin(60°)$

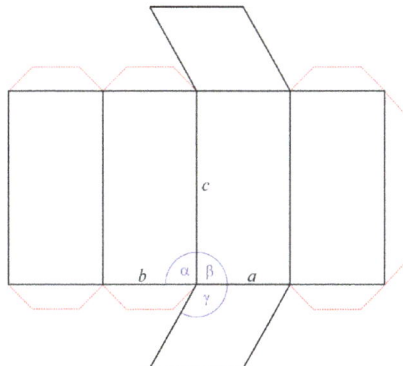

Bravais Lattice

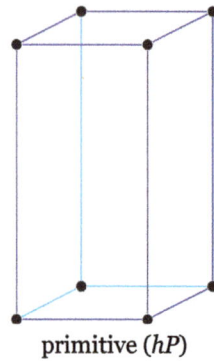

primitive (*hP*)

Trigonal Crystal System

Crystallographic Axes

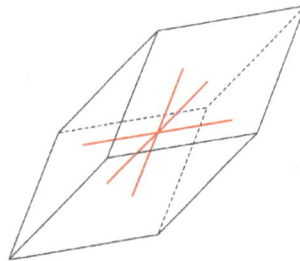

The trigonal (or rhombohedral) crystal system is described by three primitive vectors of equal length that make equal angles ($\neq 90°$) with one another. The trigonal unit cell is like a cube that has been stretched along on body diagonal.

$$a = b = c$$

$$\alpha = \beta = \gamma \neq 90°$$

An alternative cell is sometimes used to describe the trigonal crystal system. The cell is of the same shape as the conventional hexagonal unit cell with two interior points equally spaced along a diagonal.

$$a = b \neq c$$

$$\alpha = \gamma = 90° \; \beta = 120°$$

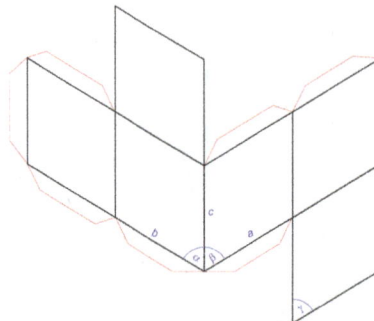

Volume of Unit Cell

Referred to rhombohedral axes

$$V = a^3 (1 - \cos\alpha) \sqrt{1 + 2\cos\alpha}$$

Referred to hexagonal axes

$$V = a^2c \sin(60°)$$

Bravais Lattices

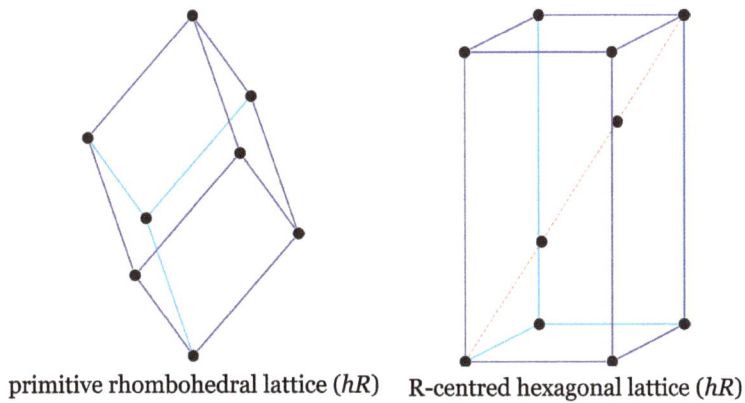

primitive rhombohedral lattice (hR)   R-centred hexagonal lattice (hR)

This figure shows the rhombohedrally-centred hexagonal cell and its relationship to the primitive rhombohedral cell.

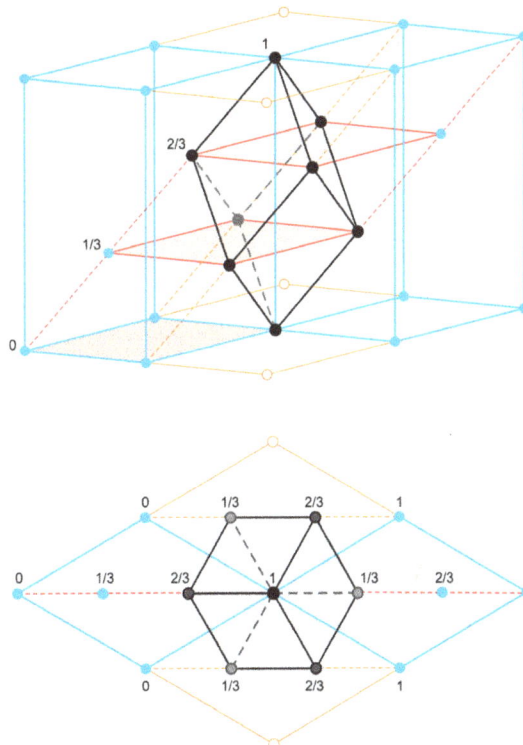

Tetragonal Crystal System

Crystallographic Axes

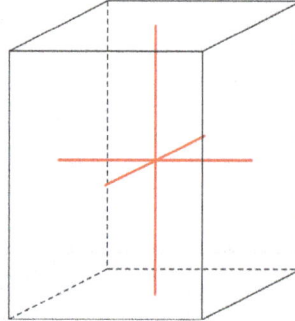

Minerals of the tetragonal crystal system are referred to three mutually perpendicular axes. The two horizontal axes are of equal length, while the vertical axis is of different length and may be either shorter or longer than the other two.

$a = b \neq c$

$\alpha = \beta = \gamma = 90°$

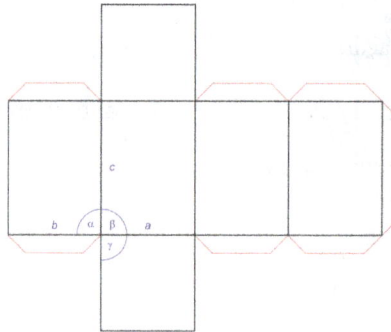

Volume of Unit Cell

$V = a^2c$

Bravais Lattices

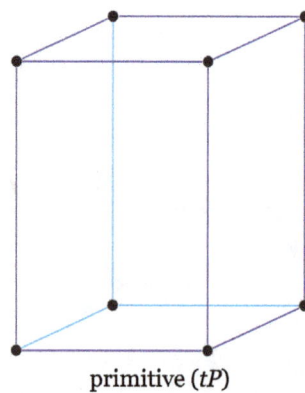

primitive (tP)

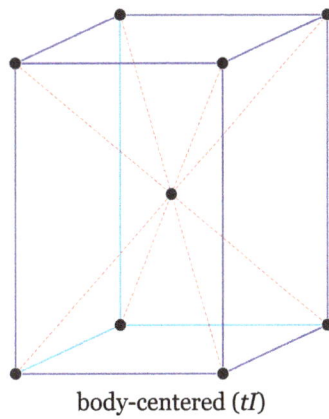

body-centered (*tI*)

Orthorhombic Crystal System

Crystallographic Axes

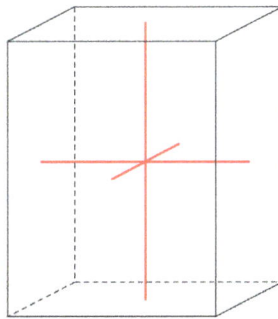

Minerals of the orthorhombic (or rhombic) crystal system are referred to three mutually perpendicular axes, each of which is of a different length than the others.

$a \neq b \neq c$

$\alpha = \beta = \gamma = 90°$

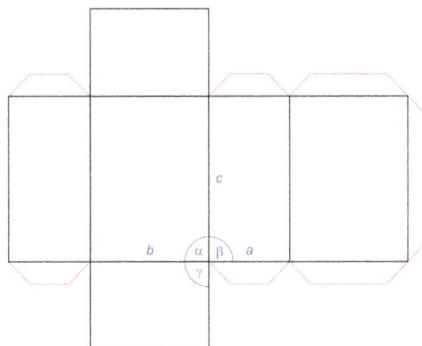

Volume of Unit Cell

$V = abc$

Bravais Lattices

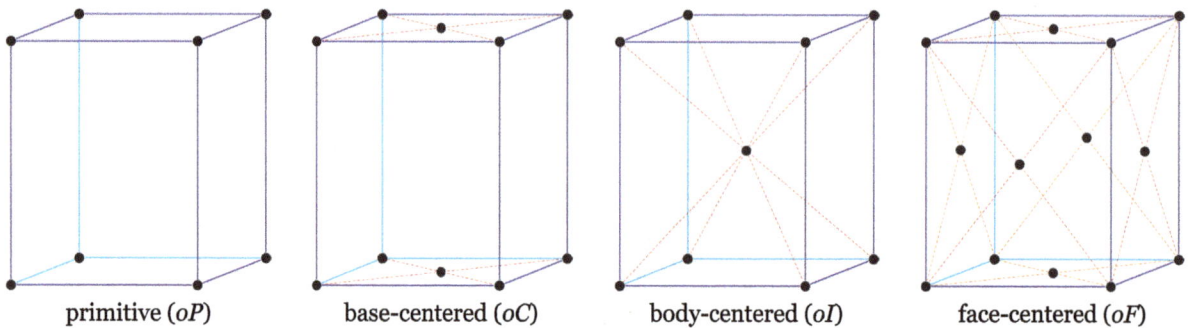

primitive (*oP*)   base-centered (*oC*)   body-centered (*oI*)   face-centered (*oF*)

Monoclinic crystal system

Crystallographic Axes

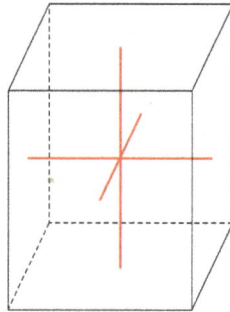

Minerals of the monoclinic crystal system are referred to three unequal axes. Two of these axes (a and c) are inclined toward each other at an oblique angle; these are usually depicted vertically. The third axis (b) is perpendicular to the other two and is called the ortho axis. The two vertical axes therefore do not intersect one another at right angles, although both are perpendicular to the horizontal axis.

$$a \neq b \neq c$$

$$\alpha = \gamma = 90° \neq \beta$$

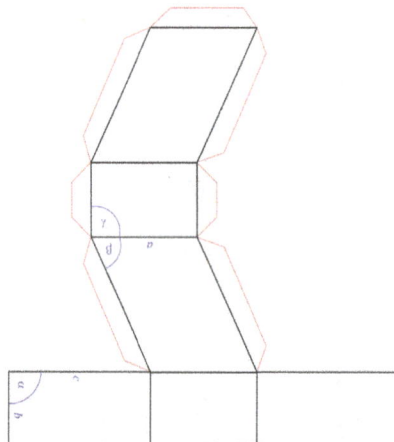

Volume of unit cell

$$V = abc \sin\beta$$

Bravais Lattices

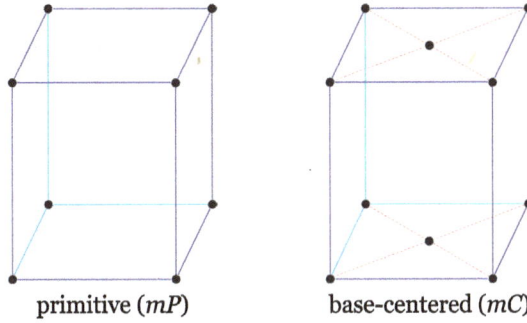

primitive (*mP*)   base-centered (*mC*)

Triclinic Crystal System

Crystallographic Axes

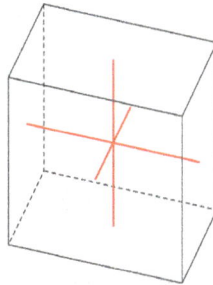

Minerals of the triclinic crystal system are referred to three unequal axes, all of which intersect at oblique angles. The triclinic system is sometimes called the anorthic system because there are no angles that are orthogonal.

$a \neq b \neq c$

$\alpha \neq \beta \neq \gamma \neq 90°$

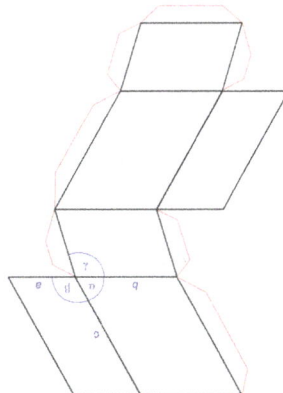

Volume of Unit Cell

$$V = abc \sqrt{1 - \cos^2\alpha - \cos^2\beta - \cos^2\gamma + 2\cos\alpha \cdot \cos\beta \cdot \cos\gamma}.$$

Bravais Lattice

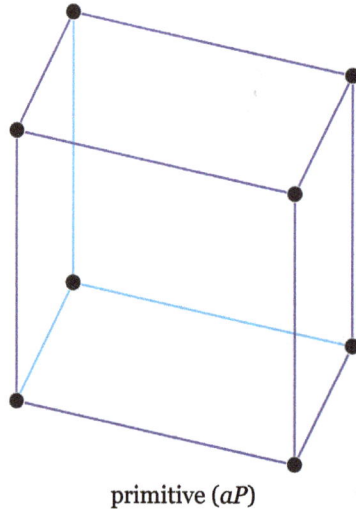

primitive (aP)

Cubic Crystal System

The cubic crystal system is defined as having the symmetry of a cube: the conventional unit cell can be rotated by 90°° about any axis, or by 180° around an axis running through the center of two opposing cube edges, or by 120° around a body diagonal, and retain the same shape. The conventional cell then takes the form

$$A_1 = a\,\hat{x}$$
$$A_2 = a\,\hat{y}$$
$$A3 = a\,\hat{z}$$

with unit cell volume

$$V = a^3.$$

This is the limiting case of both the orthorhombic and tetragonal systems when all primitive vectors are equal in length. There are three Bravais lattices in the cubic system.

Simple or primitive cubic lattice (sc or cubic-P) has one lattice point at the each corner of the unit cell. It has unit cell vectors a = b = c and interaxial angels α=β=γ=90°.

The simplest crystal structures are those in which there is only a single atom at each lattice point. In the sc structures the spheres fill 52 % of the volume. The number of atoms in a unit cell is one (8×1/8 = 1). This is only one metal (α-polonium) that have the sc lattice.

CRYSTAL LATTICE
simple cubic

$a = b = c$

$\alpha = \beta = \gamma = 90°$

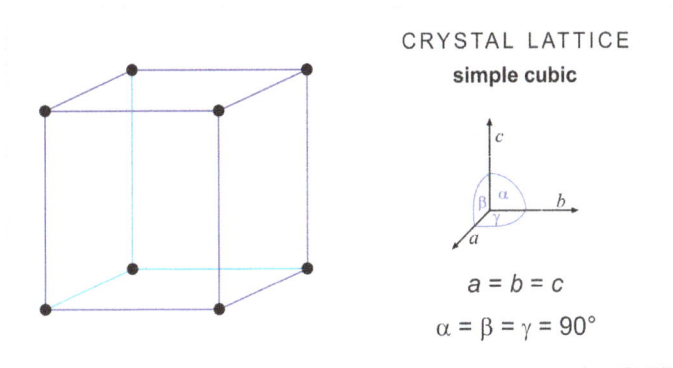

Face Centered Cubic Crystal Structure

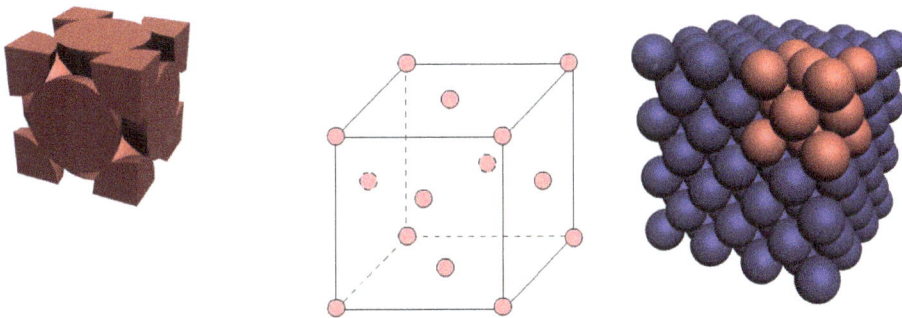

Figure's re-drawn as in the book Fundamentals of Materials Science and Engineering by William D.Callister.

Face Centered Cubic :An arrangement of atoms in crystals in which the atomic centers are disposed in space in such a way that one atom is located at each of the corners of the cube and one at the center of each face. This structure also contains the same particles in the centers of the six faces of the unit cell, for a total of 14 identical lattice points.The face-centered cubic unit cell is the simplest repeating unit in a cubic closest-packed structure.

1) Hard Sphere Unit Cell Representation. 2)A Reduced Sphere Unit Cell.3)An Aggregate of Many Atoms. In Figure all atoms are identical.

Body Centred Cubic Structure

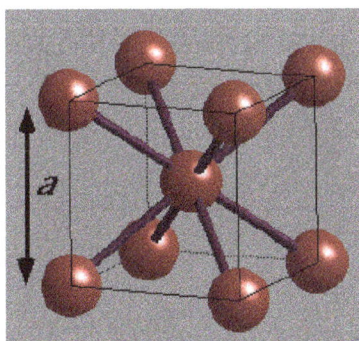

Body-Centred Cubic

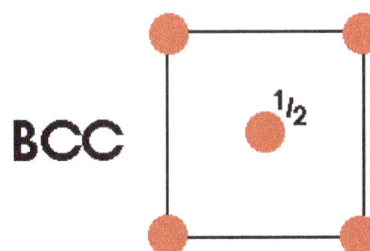

BCC

At room temperatures, elements Li, Na, K, Rb, Ba, V, Cr and Fe have structures that can be described as body centre cubic (bcc) packing of spheres. The other two common ones are face centred cubic (fcc) and hexagonal closest (hcp) packing. This type of structure is shown by the diagram above.

In a crystal structure, the arrangement extends over millions and millions of atoms, and the above diagram shows the unit cell, the smallest unit that, when repeatedly stacked together, will generate the entire structure.

Actually, the unit we draw is more than a unit cell. We use the centre of the atoms (or spheres) to represent the corners of the unit cell, and each of these atoms are shared by 8 unit cells. There is a whole atom located in the centre of the unit cell.

Usually, the length of the cell edge is represented by a. The direction from a corner of a cube to the farthest corner is called body diagonal (bd). The face diagonal (fd) is a line drawn from one vertex to the opposite corner of the same face. If the edge is a, then we have:

$$fd^2 = a^2 + a^2 = 2a^2$$
$$bd^2 = fd^2 + a^2$$
$$= a^2 + a^2 + a^2$$
$$= 3a^2$$

Atoms along the body diagonal (bd) touch each other. Thus, the body diagonal has a length that is four times the radius of the atom, R.

$$bd = 4R$$

The relationship between a and R can be worked out by the Pythagorean theorem:

$$(4R)^2 = 3a^2$$

Thus,

$$4R = sqrt(3)a$$

or

$$a = 4R / sqrt(3)$$

Recognizing these relationships enable you to calculate parameters for this type of crystal. For example, one of the parameter is the packing fraction, the fraction of volume occupied by the spheres in the structure.

Example

What is the packing fraction for a body centred packed structue?

Solution

Well, this means you should calculate the percentage of the volume occupied by the spheres in the packing.

$$Packing\ fraction\ =V_{sphere}\ /V_{unit\ cell}$$
$$= 2*4/3\pi R^3\ /\ (4/sqrt(3)R^3)$$
$$=\sqrt{3}\,\pi/8\ =\ 0.6802$$

The packing fractions are

fcc and hcp	bcc	simple cubic
74.05 %	68.01 %	52%

Hexagonal Crystal Family

Hexagonal system, one of the principal categories of structures to which a given crystalline solid can be assigned. Components of crystals in this system are located by reference to four axes—three of equal length set at 120° to one another and a fourth axis perpendicular to the plane of the other three. If the atoms or atomic groups in the solid are represented by points and the points are connected by line segments, the resulting lattice will define the edges of an orderly stacking of blocks, or unit cells. The hexagonal unit cell is distinguished by the presence of a single line, called an axis of 6-fold symmetry, about which the cell can be rotated by either 60° or 120° without changing its appearance.

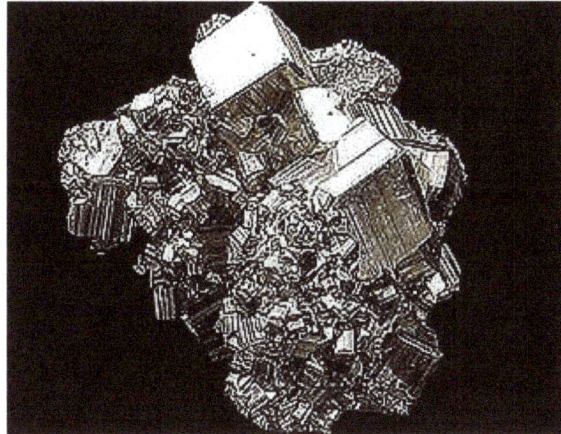

crystal systems

Minerals Crystallize according to one of seven Motifs, known as Crystal Systems.

Among the primary crystal systems, the hexagonal system has the fewest substances assigned to it, including arsenic, calcite, dolomite, quartz, apatite, tourmaline, emerald, ruby, cinnabar, and graphite.

All crystals in the hexagonal system are classed as optically uniaxial, meaning that light travels through the crystal at different speeds in different directions.

Lattice Systems

The hexagonal crystal family consists of two lattice systems: hexagonal and rhombohedral. Each lattice system consists of one Bravais lattice.

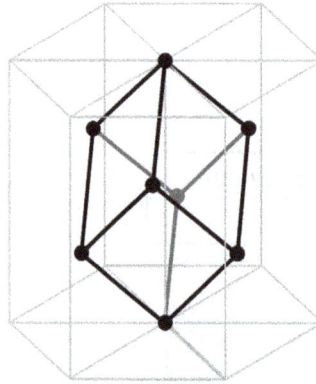

Relation between the two settings for the rhombohedral lattice

Hexagonal crystal family		
Bravais lattice	**Hexagonal**	**Rhombohedral**
Pearson symbol	**hP**	**hR**
Hexagonal unit cell		
Rhombohedral unit cell		

In the hexagonal family, the crystal is conventionally described by a right rhombic prism unit cell with two equal axes (a by a), an included angle of 120° (γ) and a height (c, which can be different from a) perpendicular to the two base axes.

The hexagonal unit cell for the rhombohedral Bravais lattice is the R-centered cell, consisting of two additional lattice points which occupy one body diagonal of the unit cell with coordinates ($\frac{2}{3}$, $\frac{1}{3}$, $\frac{1}{3}$) and ($\frac{1}{3}$, $\frac{2}{3}$, $\frac{2}{3}$). Hence, there are 3 lattice points per unit cell in total and the lattice is non-primitive.

The Bravais lattices in the hexagonal crystal family can also be described by rhombohedral axes. The unit cell is a rhombohedron (which gives the name for the rhombohedral lattice system). This is a unit cell with parameters $a = b = c$; $\alpha = \beta = \gamma \neq 90°$. In practice, the hexagonal description is more commonly used because it is easier to deal with a coordinate system with two 90° angles. However, the rhombohedral axes are often shown (for the rhombohedral lattice) in textbooks because this cell reveals $3m$ symmetry of crystal lattice.

The rhombohedral unit cell for the hexagonal Bravais lattice is the D-centered cell, consisting of two additional lattice points which occupy one body diagonal of the unit cell with coordinates ($\frac{1}{3}$, $\frac{1}{3}$, $\frac{1}{3}$) and ($\frac{2}{3}$, $\frac{2}{3}$, $\frac{2}{3}$). However, such a description is rarely used.

Crystal Systems

Crystal system	Required symmetries of point group	Point groups	Space groups	Lattice system
Trigonal	1 threefold axis of rotation	5	7	Rhombohedral
			18	Hexagonal
Hexagonal	1 sixfold axis of rotation	7	27	

The hexagonal crystal family consists of two crystal systems: trigonal and hexagonal. A crystal system is a set of point groups in which the point groups themselves and their corresponding space groups are assigned to a lattice system.

The trigonal crystal system consists of the 5 point groups that have a single three-fold rotation axis. These 5 point groups (space groups 143 to 167) have 7 corresponding space groups (denoted by R) assigned to the rhombohedral lattice system and 18 corresponding space groups (denoted by P) assigned to the hexagonal lattice system.

The hexagonal crystal system consists of the 7 point groups that have a single six-fold rotation axis. These 7 point groups have 27 space groups (168 to 194), all of which are assigned to the hexagonal lattice system. Graphite is an example of a crystal that crystallizes in the hexagonal crystal system.

Crystal Classes

Trigonal Crystal System

The trigonal crystal system is the only crystal system whose point groups have more than one lattice system associated with their space groups: the hexagonal and rhombohedral lattices both appear.

The 5 point groups in this crystal system are listed below, with their international number and notation, their space groups in name and example crystals. (All these point groups are also associated with some space groups not in the rhombohedral lattice system.)

Space group no.	Point group					Type	Examples	Space groups	
	Name	Intl	Schoen.	Orb.	Cox.			Hexagonal	Rhombo-hedral
143–146	Trigonal pyramidal	3	C_3	33	+	enantiomorphic polar	carlinite, jarosite	P3, P3$_1$, P3$_2$	R3

147–148	Rhombohedral	3	C_{3i} (S_6)	3×	$[2^+,6^+]$	centrosymmetric	dolomite, ilmenite	P3	R3
149–155	Trigonal trapezohedral	32	D_3	223	$[2,3]^+$	enantiomorphic	abhurite, alpha-quartz (152, 154), cinnabar	P312, P321, P3$_1$12, P3$_1$21, P3$_2$12, P3$_2$21	R32
156–161	Ditrigonal pyramidal	3m	C_{3v}	*33		polar	schorl, cerite, tourmaline, alunite, lithium tantalate	P3m1, P31m, P3c1, P31c	R3m, R3c
162–167	Ditrigonal scalenohedral	3m	D_{3d}	2*3	$[2^+,6]$	centrosymmetric	antimony, hematite, corundum, calcite, bismuth	P31m, P31c, P3m1, P3c1	R3m, R3c

Hexagonal Crystal System

The point groups (*crystal classes*) in this crystal system are listed below, followed by their representations in Hermann–Mauguin or international notation and Schoenflies notation, and mineral examples, if they exist.

Space group no.	Point group					Type	Examples	Space groups
	Name	Intl	Schoen.	Orb.	Cox.			
168–173	Hexagonal pyramidal	6	C_6	66	+	enantiomorphic polar	nepheline, cancrinite	P6, P6$_1$, P6$_5$, P6$_2$, P6$_4$, P6$_3$
174	Trigonal dipyramidal	6	C_{3h}	3*	$[2,3^+]$		laurelite and boric acid	P6
175–176	Hexagonal dipyramidal	6/m	C_{6h}	6*	$[2,6^+]$	centrosymmetric	apatite, vanadinite	P6/m, P6$_3$/m
177–182	Hexagonal trapezohedral	622	D_6	226	$[2,6]^+$	enantiomorphic	kalsilite and high quartz	P622, P6$_1$22, P6$_5$22, P6$_2$22, P6$_4$22, P6$_3$22
183–186	Dihexagonal pyramidal	6mm	C_{6v}	*66		polar	greenockite, wurtzite	P6mm, P6cc, P6$_3$cm, P6$_3$mc
187–190	Ditrigonal dipyramidal	6m2	D_{3h}	*223	$[2,3]$		benitoite	P6m2, P6c2, P62m, P62c
191–194	Dihexagonal dipyramidal	6/mmm	D_{6h}	*226	$[2,6]$	centrosymmetric	beryl	P6/mmm, P6/mcc, P6$_3$/mcm, P6$_3$/mmc

Hexagonal Close Packed

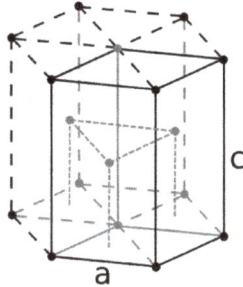

Hexagonal close packed (hcp) unit cell

Hexagonal close packed (hcp) is one of the two simple types of atomic packing with the highest density, the other being the face centered cubic (fcc). However, unlike the fcc, it is not a Bravais lattice as there are two nonequivalent sets of lattice points. Instead, it can be constructed from the hexagonal Bravais lattice by using a two atom motif (the additional atom at about $(\frac{2}{3}, \frac{1}{3}, \frac{1}{2})$) associated with each lattice point.

Example: Quartz

Quartz mineral embedded in limestone (top right of the sample),
easily identifiable by its hexagonal form.

Quartz is a crystal that belongs to the hexagonal lattice system but exists in two polymorphs that are in two different crystal systems. The crystal structures of α-quartz are described by two of the 18 space groups (152 and 154) associated with the trigonal crystal system, while the crystal structures of β-quartz are described by two of the 27 space groups (180 and 181) associated with the hexagonal crystal system.

Rhombohedral Lattice Angle

The lattice angles and the lengths of the lattice vectors are all the same for both the cubic and rhombohedral lattice systems. The lattice angles for simple cubic, face-centered cubic, and body-centered cubic lattices are $\pi/2$ radians, $\pi/3$ radians, and arccos(-1/3) radians, respectively. A rhombohedral lattice will result from lattice angles other than these.

Trigonal Crystal System

Trigonal system, also called rhombohedral system, one of the structural categories to which crystalline solids can be assigned. The trigonal system is sometimes considered to be a subdivision of the hexagonal system.

The seven primitive crystal systems

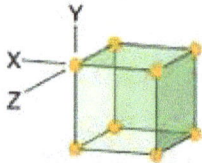

Isometric (or cubic)
All three axes are equal in length, and all are perpendicular to one another.

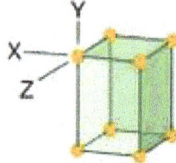

Tetragonal
Two of the three axes are equal in length, and all three axes are perpendicular to one another.

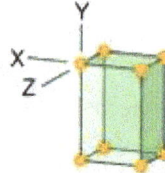

Orthorhombic
All three axes are unequal in length, and all are perpendicular to one another.

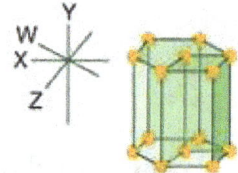

Hexagonal
Of four axes, three are of equal length, are separated by equal angles, and lie in the same plane. The fourth axis is perpendicular to the plane of the other three axes. Hexagonal cells have lattice points in each of the two six-sided faces.

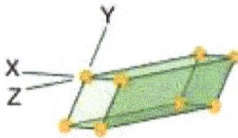

Triclinic
All three axes are unequal in length, and none is perpendicular to another.

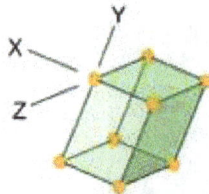

Monoclinic
All three axes are unequal in length, and two axes are perpendicular to each other.

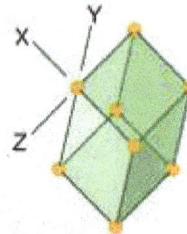

Rhombohedral (or trigonal)*
All three axes are of equal length, and none of the axes is perpendicular to another, but the crystal faces all have the same size and shape.

*Some sources do not separate the hexagonal and rhombohedral (trigonal) systems.

crystal systems;The seven primitive crystal systems

Components of crystals in the trigonal system, like those of the hexagonal system, are located by reference to four axes—three of equal length with 120° intersections and one perpendicular to the plane of the other three. The trigonal unit cell is distinguished by the presence of a single line called an axis of three-fold symmetry about which the cell can be rotated by 120° to produce a face indistinguishable from the face presented in the starting position. Selenium and other elements may crystallize in trigonal form.

Tetragonal Crystal System

Tetragonal system, one of the structural categories to which crystalline solids can be assigned. Crystals in this system are referred to three mutually perpendicular axes, two of which are equal in length.

If the atoms or atom groups in the solid are represented by points and the points are connected, the resulting lattice will consist of an orderly stacking of blocks, or unit cells. The tetragonal unit cell is distinguished by an axis of fourfold symmetry, about which a rotation of the cell through an angle of 90° brings the atoms into coincidence with their initial positions. The elements boron and tin can crystallize in tetragonal form, as can some minerals such as zircon.

Tetragonal crystal system

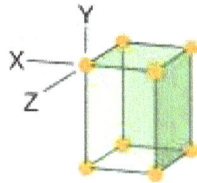

Two of the three axes are equal in length, and all three axes are perpendicular to one another.

Its one variant is:

Body-centred tetragonal
lattice point in the middle of the unit cell

tetragonal crystal system

Crystals in a tetragonal system are characterized by three mutually perpendicular axes, two of which are equal in length.

The Ditetragonal-bipyramidal Class

(= Holohedric Division)   4/m 2/m 2/m

In the next Figure we depict the tetragonal system of crystallographic axes plus a derived proto-pyramid.

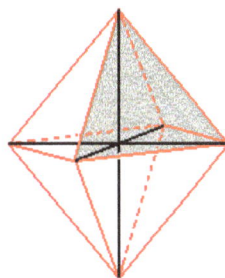

m **P.**

Tetragonal system of crystallographic axes, and a derived Protopyramid
as a possible Form of the Class. One face of this Form is emphasized by a grey shade.

The symmetry content (bundle) of the Ditetragonal-bipyramidal Crystal Class is :

* Five mirror planes, namely one main mirror plane, lying in the plane of the secondary axes, and four secondary mirror planes which are perpendicular to the main mirror plane and intersect in the vertical crystallographical axis, while having angles of 45° between each other. Two of these mirror planes are equivalent and go through the horizontal crystallographic

axes (i.e. each one of them contains the vertical axis and one of the horizontal axes), the two other equivalent mirror planes go right between the former.

- Five rotation axes, namely one 4-fold main rotation axis, coincident with the crystallographic main axis, and perpendicular to the main mirror plane, and two + two 2-fold rotation axes, two of them coincident with the horizontal crystallographical axes, the two others lying in between them.

- Center of symmetry.

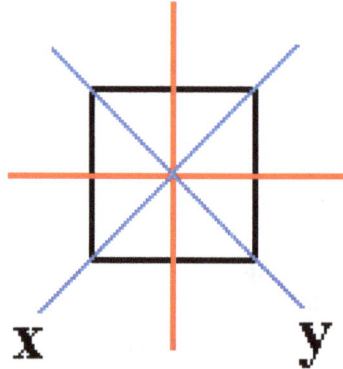

The equatorial plane of the tetragonal protopyramid, and the traces of the 2 + 2 vertical mirror planes. One set of them (blue) contains, besides the vertical, the horizontal crystallographic axes, the other set makes an angle of 45° with the first set, and only contains the vertical crystallographic axis.

The Holohedric Division of the Tetragonal Crystal System has seven primary Forms :

- Protopyramid.

- Deuteropyramid.

- Ditetragonal pyramid.

- Protoprism.

- Deuteroprism.

- Ditetragonal prism.

- Basic Pinacoid.

(The pyramids are in fact bipyramids) We shall discuss them all.

There are seven basic types of faces, and when we subject each face to the symmetry of our present Crystal Class (which means that we demand that the Forms generated from the faces have the symmetry of the Class) we generate the above Forms. The faces are :

- (2) a : a : c

- (3) a : ~a : c

- (1) a : na : mc

- (5) a : a : ~c

- (6)   a : ~a : ~c

- (4)   a : na : ~c

- (7)   ~a : ~a : c

The numbers (1), (2), etc. correspond to the numbered positions of these faces in the stereographic projection of Figure.

Besides the Forms of this Class, also the Forms of all the other Classes of the Tetragonal Crystal System can be derived, by subjecting these faces to the symmetries of the relevant Class. This I will later call the facial approach for deriving the Forms. The Forms of the lower symmetry Classes can also be derived by means of the concepts of *Holohedric, Hemihedric,* etc., which I will call the merohedric approach. Of course for the derivation of Holohedric Forms we must resume to the facial approach. We can make use of the stereographic projection of faces (face poles) and symmetry elements (like rotation axes and mirror planes).

Figure depicts a stereogram (i.e. a stereographic projection) of the symmetry elements of the Holohedric Class, and the seven basic faces compatible with the Tetragonal Crystal System (indicated by the numbers corresponding with those of the above listing of these faces) :

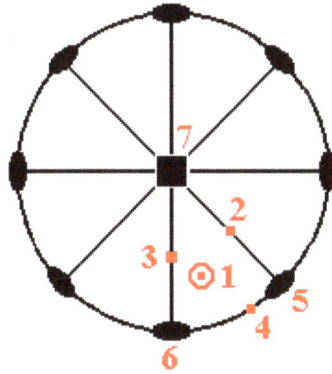

Stereographic projection of the symmetry elements of the Ditetragonal-bipyramidal Crystal Class (= Holohedric Division), and of the seven basic faces compatible with the Tetragonal Crystal System.

The indicated faces are multiplied by the symmetry elements and accordingly show up in the stereograms. But in the present figure only one representation of each face is shown.

In the stereogram of Figure. we see a (vertical) 4-fold rotation axis, represented by a black square. This axis, as represented in the Figure, is perpendicular to the screen (or paper). Perpendicular to this axis is the equatorial mirror plane, represented by a solid circle in the plane of the screen (paper). It coincides with the circumference of the projection plane. Further we see (the traces of) two sets of vertical mirror planes. One set of which contains, besides the main axis, the secondary crystallographic axes, i.e. one member of this set contains one secondary axis, the other member contains the other. The second set of vertical mirror planes bisects the angles obtaining between the planes of the first set. All these vertical mirror planes are represented by straight solid lines. Finally we see two sets of 2-fold rotation axes. One set coinciding with the secondary crystallographic axes, the other set bisects the angles between the members of the first set. They are represented by small solid ellipses. So all the symmetry elements of the Class are represented in the stereogram.

The seven basic faces, compatible with the Tetragonal Crystal System are also represented in the Figure. As has been said, the numbers correspond to those of the above list of basic faces :

- 1 is the most general face, a : na : mc. It is inclined in two directions, expressed by the fact that in the stereographic projection it lies within the projection plane. It can become a face of a ditetragonal pyramid. It is represented by a dot. I have also indicated its reflectional counterpart (in virtue of the horizontal mirror plane), namely as a little circle. Its projection coincides with its reflectional counterpart [For the other faces -- represented as points -- (2, 3, 4, 5, 6 and 7) I have not indicated their possible counterparts (in the vertical direction), and, moreover, the vertical faces, which appear as points on the circumference of the projection plane do not have reflectional counterparts in the vertical direction anyway].

- 2 is the face a : a : c. It is the unit face of the Tetragonal Crystal System. It can become a face of a protopyramid.

- 3 is the face a : ~a : c. It can become a face of a deuteropyramid.

- 4 is the face a : na : ~c. It can become a face of a ditetragonal prism.

- 5 is the face a : a : ~c. It can become a face of a protoprism.

- 6 is the face a : ~a : ~c. It can become a face of a deuteroprism.

- 7 (in the center of the Figure)is the face ~a : ~a : c. It can become a face of a basic pinacoid.

Now it is easy to derive the Forms of the present Crystal Class (= the Holohedric Division of the Tetragonal System) :

The primary protopyramid, primary bipyramid of type I, is derived from the upper right front face (the unit face of the Tetragonal Crystal System) -- a : a : c, number 2 of Figure -- of the image in Figure (below) -- considered as just being present on its own, without other faces, and which is in fact a certain conspicuous face on some tetragonal crystal chosen to serve as a unit face. The primary protopyramid is derived from this face by demanding that the generated Form possesses the complete symmetry bundle of the present Class : The face is multiplied by the 4-fold rotation axis resulting in a total of four faces. Then this quadruplet will be doubled by the horizontal mirror plane resulting in the protopyramid.

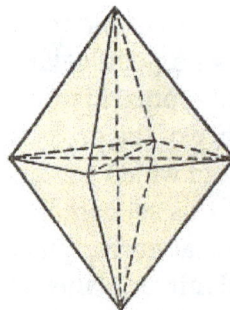

PRIMARY PROTOPYRAMID **P.**
The primary Tetragonal Protopyramid (= Tetragonal Bipyramid of type I).

Each face of this Form cuts off a finite piece of each crystallographic axis, and, considering it as the basic Form we can denote it by the Weiss symbol (a : a : c), in which the two a's express the equivalence of the two horizontal crystallographic axes, and the one c expresses the fact that the vertical crystallographic axis is not equivalent to either horizontal one.

The protopyramid is a bipyramid bounded by eight isosceles triangles. Its equatorial plane has the shape of a square. The Naumann symbol is P, and the Miller symbol is {111} (the derivation coefficients of the Weiss symbol and the indices of the Miller symbol are set equal to 1, because all other Forms are considered to be derivations of this one Form).

The stereographic projection of the (faces, i.e. all the faces, of the) tetragonal protopyramid is depicted in Figure.

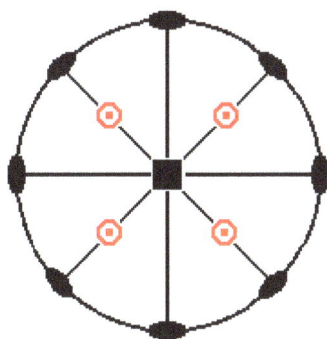

Stereogram of the Tetragonal Protopyramid. An upper face (pole) is represented by a
red dot. A lower face (pole) is represented by a small circle. In the present case their positions
on the projection plane coincide. So we see four sets of two faces each, an upper and a lower one.

The first derived Forms are bipyramids like the protopyramid but with a different a : c ratio when compared with the primary protopyramid of Figure. The general Weiss symbol for such pyramids is consequently (a : a : mc), the Naumann symbol is mP (the letter before P relates to the cut-off piece of the vertical crystallographic axis), the Miller symbol is {hhl}.

In figure we already saw such a derived protopyramid (assuming m having there some rational value), and in the figure below we see two examples of derived protopyramids with m = 2 and m = 1/2respectively (If we would set up a stereogram of such a derived protopyramid we would get the same picture as Figure, but with the projections of the faces closer to the perifery of the projection plane or closer to the center of it respectively.

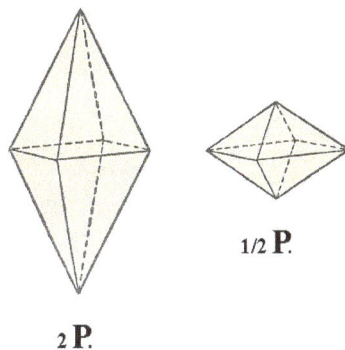

1/2 **P.**

2 **P.**

DERIVED PROTOPYRAMIDS

Derived Tetragonal Protopyramids, with m = 2 (left) and m = 1/2 (right).

In the next Figure we see two more Forms, namely (1) the deuteropyramid and the ditetragonal pyramid.

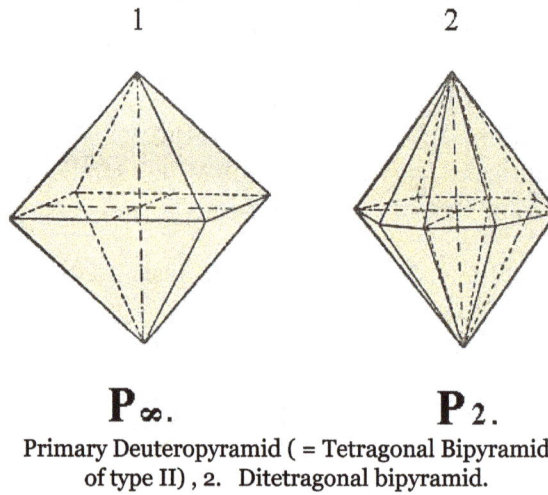

P∞.   **P₂.**

Primary Deuteropyramid (= Tetragonal Bipyramid
of type II) , 2. Ditetragonal bipyramid.

While a protopyramid is generated by subjecting the face a : a : mc to all the symmetry operations of the present Crystal Class (when m = 1 we get the primary protopyramid), a deuteropyramid is generated by subjecting the face a : ~a : mc to all those symmetry operations (when m = 1 -- 3 in Figure -- we get the primary deuteropyramid). The shape of the deuteropyramid (tetragonal bipyramid of type II) does not differ from that of the protopyramid. It differs in its orientation with respect to the crystallographic axes. Each of its faces cuts off pieces only from the main axis and from one of the secondary (i.e. horizontal) axes, and is parallel to the other secondary axis. So the Weiss symbol is (a : ~a : c), or generally, because also derived pyramids are possible, (a : ~a : mc), and the Naumann symbol is mP~ (the sign before P relates to the main axis, the sign after P to the secondary axis)(Recall that the sign ~ stands for infinity, in Figures denoted by a horizontal 8). The Miller symbol is {hol}.

The stereographic projection of the deuteropyramid is depicted in Figure.

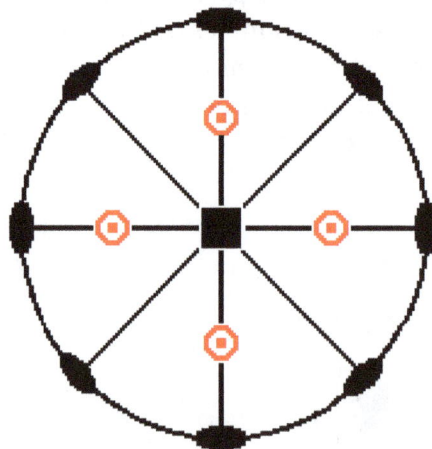

Stereogram of the Tetragonal Deuteropyramid.

When we subject the face a : na : mc, i.e. a face of which the orientation with respect to the crystallographic axes is the most general, to all the symmetries of the present Crystal Class, we obtain a ditetragonal bipyramid, as yet another Form of the Holohedric Division. The ditetragonal bipyramid has 16 faces, which are oriented such that each of them intersects all three axes at different distances. The equatorial plane is a ditetragon, i.e. an octagon (= figure with eight sides) having all sides of equal length but with alternating equal angles.

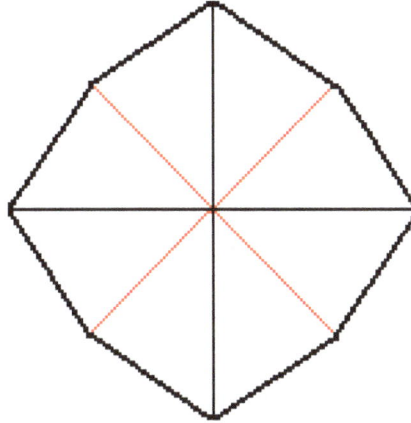

A Ditetragon, the equatorial plane of the Ditetragonal bipyramid.

The general Weiss symbol is (a : na : mc), the Naumann symbol is mPn, and the Miller symbol is {hkl}. Limiting Forms are the protopyramid, when n = 1, and the deuteropyramid, when n = ~.

The stereographic projection of the ditetragonal bipyramid is depicted in Figure.

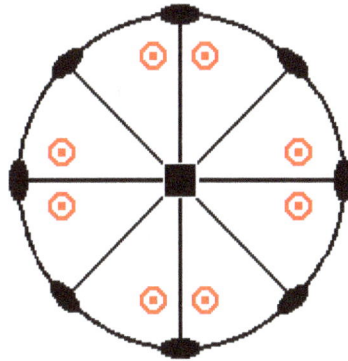

Stereogram of the Ditetragonal Bipyramid.

When for the described pyramids the derivation coefficient m increases the pyramids become more and more sharp. When the coefficient finally is equal to infinity the sides of the pyramids have become vertical (i.e. parallel to the main crystallographic axis) and the Forms become prisms. These are open Forms which, in their conventional oriention do not enclose space completely. Their top and bottom are open.

From the (primary) protopyramid we derive the protoprism (= tetragonal prism of type I), by letting the unit face, the one which is situated in front, up, and to the right, become steeper and steeper until it is vertically oriented, implying that the cut-off piece (intercept) of the vertical axis is infinite (in length), while the cut-offs of the other two axes remain the same. This new face can be denoted by a : a : ~c. When we subject this face to all the symmetries of our Crystal Class we will obtain a protoprism as a new Form.

Its Weiss symbol is (a : a : ~c), the Naumann symbol is ~P., and the Miller symbol is {110}.

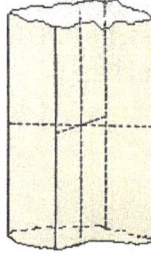

PROTOPRISM   ∞ **P.**

A Protoprism, derived from a (face of the) Protopyramid. An open Form.

Because the protoprism (and also all other prisms) is an OPEN Form, the next Figure does not correctly depict it, by reason of its suggesting the presence of top and bottom faces (while the protoprism lacks these faces). In fact the Figure depicts a combination of protoprism and basic pinacoid.

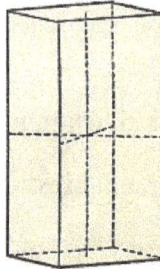

PROTOPRISM

+

BASIC PINACOID

A combination of a Protoprism and a Basic Pinacoid,
forming the top and bottom faces of the (combined) Form.

The stereographic projection of the protoprism is depicted in Figure.

Stereogram of the Tetragonal Protoprism.

The projection of each face is indicated in the figure as if they represented the coincidence of an upper and lower face pole. But in fact each prism face leads to only one face pole. I depict them nevertheless as is done in the figure in order that the symbols for the 2-fold rotation axes are not obscured by the (projections of the) poles.

A next Form is the deuteroprism (= tetragonal prism of type II). It derives from the deuteropyramid in the same way that the protoprism was derived from (a face of) the protopyramid, and is distinguished from the protoprism only by its orientation with respect to the crystallographic axes. So the deuteroprism is generated from the face a : ~a : ~c.

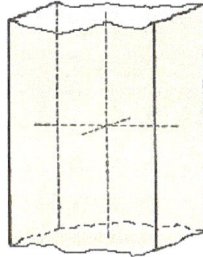

$$\infty \, \mathbf{P} \infty.$$

DEUTEROPRISM
Deuteroprism, an open Form of the Holohedric Division.

The following Figure gives this same deuteroprism, but now closed by a basic pinacoid :

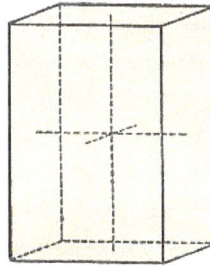

DEUTEROPRISM
+
BASIC PINACOID
A Deuteroprism, closed by a basic (i.e. horizontal) Pinacoid.

The Weiss symbol for the deuteroprism is (a : ~a : ~c), the Naumann symbol is ~P~., and the Miller symbol is {100}.

The stereographic projection of the deuteroprism is depicted in Figure.

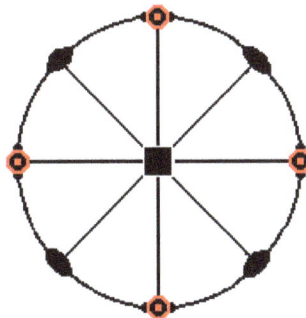

Stereographic projection of the Tetragonal Deuteroprism.
Concerning the representation of the projection of the face poles.

The next Form is the ditetragonal prism. It can be derived from the ditetragonal bipyramid (Figure, by isolating a face and letting it become vertical. We then have the face a : na : ~c. When we subject this face to all the symmetries of our Crystall Class we'll get the ditetragonal prism: The isolated face will be doubled by the nearby mirror plane resulting in two symmetrically related faces having an angle between each other smaller than 180°. This pair of faces will then be multiplied four times by the action of the 4-fold rotation axis, resulting in the ditetragonal prism, consisting of eight vertical faces. Its horizontal outline (i.e. when seen from above, or below) is the same as that of the equatorial plane of the corresponding ditetragonal bipyamid.

The Weiss symbol is (a : na : ~c), its Naumann symbol is ~Pn., and the Miller symbol is {hko}.

Here we depict the ditetragonal prism as if it is closed by a basic pinacoid (top and bottom faces) for reasons of clarity.

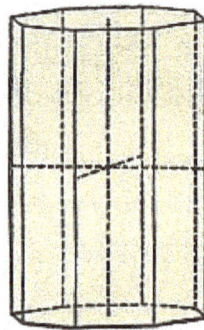

$$\infty \mathbf{P}_n.$$

The Ditetragonal Prism.

The stereographic projection of the ditetragonal prism is depicted in Figure.

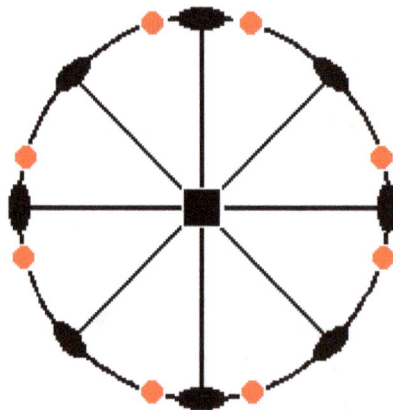

Stereogram of the Ditetragonal Prism.

Finally we have the basic pinacoid. We already spoke about it earlier. Like the prisms it is an open Form. It consists of two horizontal parallel faces (because the faces are horizontal it is called a basicpinacoid. In other Crystal Classes there also exist pinacoids which are not oriented horizontally). A pinacoid can also be called parallelohedron.

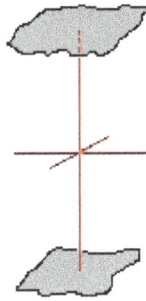

BASIC PINACOID   0 **P.**
A Basic Pinacoid (Basic Parallelohedron), another open
Form of the Holohedric Division.

The basic pinacoid will be generated when we take a horizontally oriented face -- such a face can be denoted by ~a : ~a : c -- and subject it to the symmetry of the present Class. The 4-fold rotation axis and the vertical mirror planes will have no effect, but the horizontal mirror plane will duplicate the face (the 2-fold rotation axes will have the same effect), so we'll end up with a Form consisting of two parallel horizontal faces, the basic pinacoid, which is, as has been said, an open Form and can only occur in certain combinations, for example with a prism.

The Weiss symbol is (~a : ~a : c), the Naumann symbol is 0P., and the Miller symbol is {001}.

The "0" in the Naumann symbol 0P means that the face is inclined by an amount of zero, i.e. it is horizontal. We can think of it as the upper right front face of the protopyramid, with its steepness decreased all the way down to zero, i.e. its coefficient for c has become zero. But of course the coefficient can have whatever value, including 1, because varying its hight above or below the origin of the system of crystallographic axes does not change the crystallographic character of the face (a parallel shift of a face does not change its crystallographic character, but notice that parallel faces are not necessarily crystallographically equivalent. The two faces of a pinacoid are equivalent, they involve the same atomic aspect to the environment, and so implying the same physical behavior. In fact all the faces constituting a Form are equivalent in this respect. There exist (in other Classes than the present one) Forms consisting of just one face. Such a Form is called a pedion. And it is possible that, together with other Forms, two pedions combine, forming a pair of parallel faces. But these faces are not equivalent because they belong to different forms, a pedion and another pedion).

The stereographic projection of the basic pinacoid is depicted in Figure.

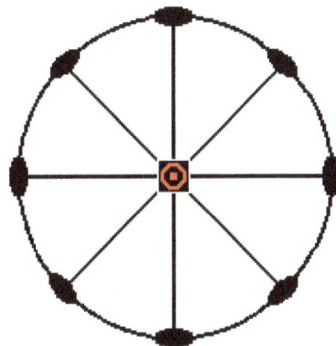

Stereogram of the Basic Pinacoid.

All these Forms can engage in combinations with each other. In the Figures we show some examples.

1   2

$$\infty P . P \qquad \infty P \infty . P$$

1. Protoprism + Protopyramid. 2. Deuteroprism + Protopyramid.

1   2

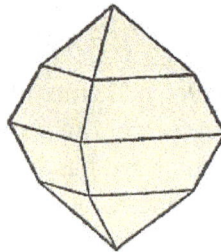
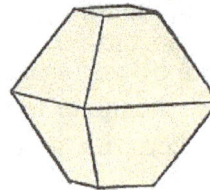

$$P . \tfrac{1}{m} P \qquad P . {}_0 P$$

Protopyramid + derived Protopyramid. 2. Protopyramid + Basic Pinacoid (The polar corners of the pyramid are cut off by the Basic Pinacoid).

Bravais Lattices

Two-dimensional

There is only one tetragonal Bravais lattice in two dimensions: the square lattice.

Three-dimensional

There are two tetragonal Bravais lattices: the simple tetragonal (from stretching the simple-cubic lattice) and the centered tetragonal (from stretching either the face-centered or the body-centered cubic lattice). One might suppose stretching face-centered cubic would result in face-centered tetragonal, but face-centered tetragonal is equivalent to body-centered tetragonal, BCT (with a smaller lattice spacing). BCT is considered more fundamental, so that is the standard terminology.

Bravais lattice	Primitive tetragonal	Body-centered tetragonal
Pearson symbol	tP	tI
Unit cell		

Crystal classes

The point groups that fall under this crystal system are listed below, followed by their representations in international notation, Schoenflies notation, orbifold notation, Coxeter notation and mineral examples.

#	Point group					Type	Example	Space groups	
	Name	Intl	Schoen.	Orb.	Cox.			Primitive	Body-centered
75–80	Tetragonal pyramidal	4	C_4	44	+	enantiomorphic polar	pinnoite, piypite	P4, P4$_1$, P4$_2$, P4$_3$	I4, I4$_1$
81–82	Tetragonal disphenoidal	4	S_4	2×	[2$^+$,4$^+$]		cahnite, tugtupite	P4	I4
83–88	Tetragonal dipyramidal	4/m	C_{4h}	4*	[2,4$^+$]	centrosymmetric	scheelite, wulfenite, leucite	P4/m, P4$_2$/m, P4/n, P4$_2$/n	I4/m, I4$_1$/a
89–98	Tetragonal trapezohedral	422	D_4	224	[2,4]$^+$	enantiomorphic	cristobalite, wardite	P422, P42$_1$2, P4$_1$22, P4$_1$2$_1$2, P4$_2$22, P4$_2$2$_1$2, P4$_3$22, P4$_3$2$_1$2	I422, I4$_1$22
99–110	Ditetragonal pyramidal	4mm	C_{4v}	*44		polar	diaboleite	P4mm, P4bm, P4$_2$cm, P4$_2$nm, P4cc, P4nc, P4$_2$mc, P4$_2$bc	I4mm, I4cm, I4$_1$md, I4$_1$cd
111–122	Tetragonal scalenohedral	42m	D_{2d} (V_d)	2*2	[2$^+$,4]		chalcopyrite, stannite	P42m, P42c, P42$_1$m, P42$_1$c, P4m2, P4c2, P4b2, P4n2	I4m2, I4c2, I42m, I42d

| 123–142 | Ditetragonal dipyramidal | 4/mmm | D_{4h} | *224 | [2,4] | centrosymmetric | rutile, pyrolusite, zircon | P4/mmm, P4/mcc, P4/nbm, P4/nnc, P4/mbm, P4/mnc, P4/nmm, P4/ncc, P4$_2$/mmc, P4$_2$/mcm, P4$_2$/nbc, P4$_2$/nnm, P4$_2$/mbc, P4$_2$/mnm, P4$_2$/nmc, P4$_2$/ncm | I4/mmm, I4/mcm, I4$_1$/amd, I4$_1$/acd |

Orthorhombic Crystal System

Orthorhombic system, one of the structural categories systems to which crystalline solids can be assigned. Crystals in this system are referred to three mutually perpendicular axes that are unequal in length.

Orthorhombic system ;Orthorhombic system of Aragonite.

If the atoms or atom groups in the solid are represented by points and the points are connected, the resulting lattice will consist of an orderly stacking of blocks, or unit cells. The orthorhombic unit cell is distinguished by three lines called axes of twofold symmetry about which the cell can be rotated by 180° without changing its appearance. This characteristic requires that the angles between any two edges of the unit cell be right angles but the edges may be any length. Alpha-sulphur, cementite, olivine, aragonite, orthoenstatite, topaz, staurolite, barite, cerussite, marcasite, and enargite crystallize in the orthorhombic system.

Crystals in an orthorhombic system are characterized by three mutually perpendicular axes that are unequal in length.

The Basic Form of the Orthorhombic Crystal System, with its Weissian symbol (a : b : c), its Naumann symbol P, and its Miller symbol {111}.

(a refers to the brachy axis, b to the macro axis, and c to the vertical axis).

Orthorhombic crystal system

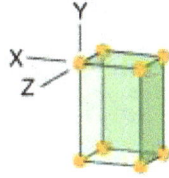

All three axes are
unequal in length, and
all are perpendicular
to one another.

Its three variants are:

Body-centred orthorhombic
lattice point in the middle of the unit cell

Base-centred orthorhombic
lattice points in the middle of each of the two ends

Face-centred orthorhombic
lattice points in the middle of each side

orthorhombic crystal system

It is a rhombic bipyramid, implying that its equatorial plane is a rhombus, not a square. This bipyramid consists of eight faces of which each is an unequilateral triangle. It has four equal middle edges and eight polar edges four of which are more obtuse while the other four are more acute. In the conventional orientation two obtuse polar edges, an upper one and a lower one, are directed to the beholder. It has six corners, and every two corners that are opposite to each other are equal. The crystallographic axes pass through the corners.

The symbols for this Basic Form, which is called a primary protopyramid are given above.

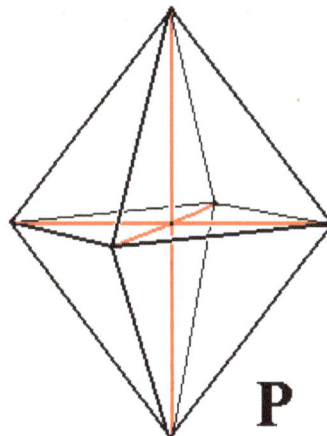

P

The Primary Protopyramid as Basic Form for the Orthorhombic Crystal System.
All the Forms of the highest symmetrical Class can be derived from this Basic Form.

The Rhombic-bipyramidal Class (= Holohedric Division) 2/m 2/m 2/m

The symmetry content of this Class is given above. It can also be depicted stereographically, as is done in the next Figure :

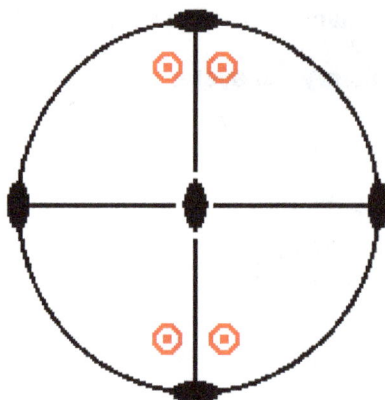

Stereographic projection of the symmetry elements of the Rhombic-bipyramidal Crystal Class, and all the faces of the most general Form, the Rhombic Bipyramid (The position of the faces -- face poles -- is meant to be a general one, as actually depicted it represents a pyramid of the Macrodiagonal Series).

Solid ellipses stand for 2-fold rotation axes. The solid circle represents a horizontal mirror plane, while the two straight diagonal lines represent two vertical mirror planes. The small red circles centred with a red dot represent upper and lower faces (face pairs), which -- as face poles -- coincide with each other on the plane of projection, i.e. the projections of the upper and the lower face of such a face pair end up at the same location on the plane of projection.

Derivation of the Forms of the Holohedric Division of the Orthorhombic Crystal System.

There are two types of Forms to be derived, closed Forms and open Forms.

Closed Forms (Pyramids)

From the above described Basic Form -- the primary (proto-)pyramid -- a number of secondary pyramids can be derived whose axes stand in a rational relationship with those of the primary pyramid. And this does not -- like it does in the Tetragonal and Hexagonal Systems -- concern only one axis, the vertical axis, but all three axes (because there is no main axis to be found in the Orthorhombic Crystal System). This implies that we have three series of derived pyramids :

- Vertical Series, Protopyramids. The members of this Series are obtained by varying the relative c axis cut-off distance.

- Brachy diagonal Series, Brachypyramids. The members of this Series are obtained by varying the relative a axis cut-off distance.

- Macro diagonal Series, Macropyramids. The members of this Series are obtained by varying the relative b axis cut-off distance.

Let us illustrate these three series.

Varying the relative c axis cut-off distance, which is equivalent to varying the derivation coefficient m in (a : b : mc) and in mP, gives us the Vertical Series, the Protopyramids which can be denoted by the Weissian symbol (a : b : mc), the Naumann symbol mP and the Miller symbol {hhl} :

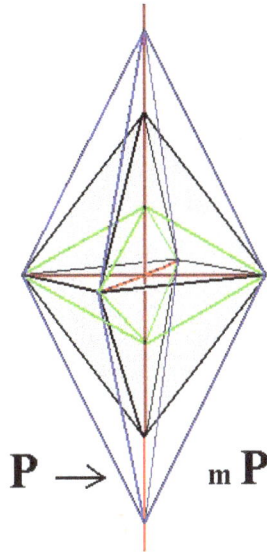

$$P \rightarrow \; _m P$$

From the Primary Pyramid (P) can be derived the Vertical Series, the Protopyramids (mP), by varying
the derivation coefficient m. Two possible non-primary (i.e. secondary) Protopyramids are indicated
(blue and green respectively). The Primary Pyramid is a Prot*O*pyramid in which m = 1.

Varying the relative a axis cut-off distance, which is equivalent to varying the derivation
coefficient n in (na : b : c) and in Pn, gives us the Brachydiagonal Series (Brachypyramids). Because
we can derive such brachypyramids from each protopyramid (including the primary protopyramid) the (general) Weissian symbol for a brachypyramid is (na : b : mc), and the corresponding
Naumann symbol is $_m \check{P}_n$. The Miller symbol is {khl}.

In the next Figure I give a brachypyramid derived from the primary protopyramid where m = 1. :

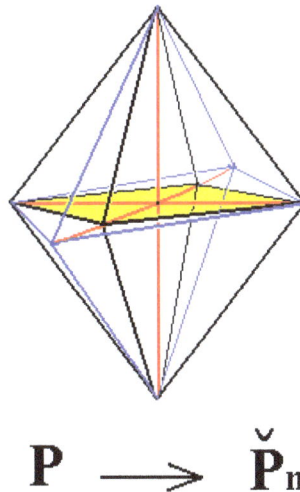

$$P \longrightarrow \check{P}_n$$

From the Primary Pyramid can be derived a Brachy diagonal Series, the primary Brachypyramids,
by varying the derivation coefficient n of the Brachy axis.

Also from any non-primary Protopyramid (i.e. a non-primary member of the Vertical Series) a
Brachypyramid can be derived in the same way, yielding the Brachy diagonal Series.

Varying the relative b axis cut-off distance, which is equivalent to varying the derivation coefficient n in (a : nb : c) and in Pn, gives us the Macrodiagonal Series (Macropyramids). Because we can derive such macropyramids from each protopyramid (including the primary protopyramid) the (general) Weissian symbol for a macropyramid is (a : nb : mc), and the corresponding Naumann symbol is $m\bar{P}n$. The Miller symbol is {hkl}.

In the next Figure I give a macropyramid derived from the primary protopyramid where m = 1. :

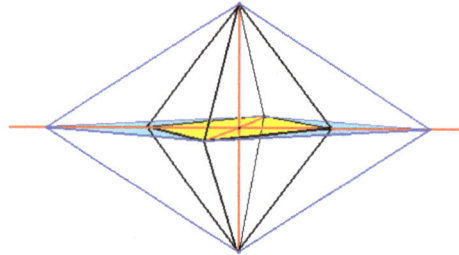

$$P \longrightarrow \bar{P}n$$

A primary Macropyramid (blue), derived from the primary Protopyramid.]
Also from any Protopyramid of the Vertical Series can a Macropyramid be derived,
by extention of the macro axis.

Open Forms (vertical and horizontal prisms and pinacoids).

From the above pyramids (protopyramids, brachypyramids, macropyramids) we can -- to begin with -- derive the corresponding vertical rhombic prisms :

- The Rhombic Protoprism.

- The Brachyprisms.

- The Macroprisms.

When we let the derivation coefficient m in the protopyramids (a : b : mc) become infinite, i.e. when we make the c axis cut-off distance of infinite length, the result will be a vertical prism, the protoprism, or, as it can be called, a primary prism, that can be denoted by (a : b : ~c) (in which "~" stands for infinity), ~P or {110}. There is only one such protoprism, its faces cutting off unit pieces from the brachy axis as well as from the macro axis, while being parallel to the vertical axis. Its horizontal section is a rhombus (this is because the unit pieces of the brachy and macro axis do not have equal lengths). The next Figure depicts the protoprism.

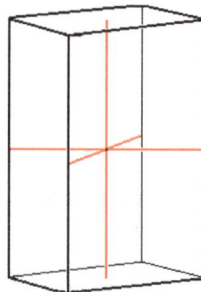

$$\infty P$$

From any Rhombic Protopyramid (a : b : mc) can be derived the
Rhombic Protoprism (a : b : ~c) by making m infinitely large.

From the pyramids of the Brachydiagonal Series (na : b : mc) a second category of vertical prisms, the brachyprisms (na : b : ~c) (Miller symbol {kh0}) can be derived when we again make the derivation coefficient m infinitely large :

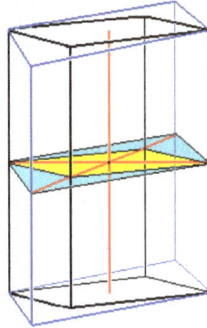

$$\infty P \implies \infty \check{P}_n$$

From any Rhombic Brachypyramid (na : b : mc) can be derived a Rhombic Brachyprism (na : b : ~c) by making m infinitely large. Here we derive such a Brachyprism directly from the Rhombic Protoprism, by extention of the cut-off distances of the brachy axis.

Like in the Rhombic Protoprism the horizontal section of the Brachyprisms is a rhombus.

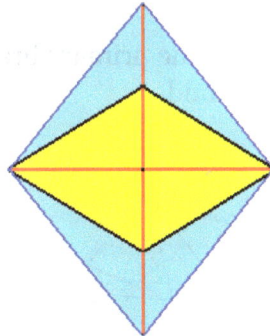

Horizontal sections through a Rhombic Protoprism (yellow) and a possible Rhombic Brachyprism (blue). Both are rhombi.

From the pyramids of the Macrodiagonal Series (a : nb : mc) a third category of vertical prisms, the macroprisms (a : nb : ~c) ({hk0}) can be derived when we again make the derivation coefficient minfinitely large :

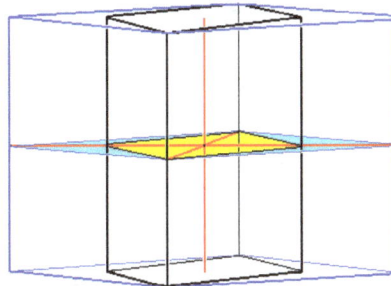

$$\infty P \implies \infty \bar{P}_n$$

From any Rhombic Macropyramid (a : nb : mc) can be derived a Rhombic Macroprism (a : nb : ~c) by making m infinitely large. Here we derive such a Macroprism directly from the Rhombic Protoprism, by extention of the cut-off distances of the macro axis.

Also the horizontal sections of the macroprisms are rhombi, as the next Figure illustrates :

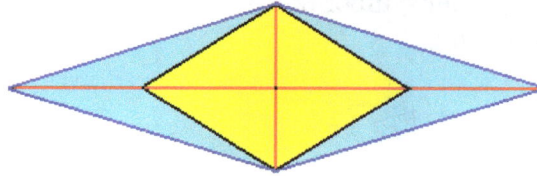

Horizontal sections through the Rhombic Protoprism (yellow) and
a possible Macroprism (blue).

Further we can derive two categories of horizontal rhombic prisms that often are also called domes :

- From the brachypyramids (na : b : mc) we can derive the corresponding brachydomes (~a : b : mc) ({ohl}) by making n infinite, which means that we get horizontal prisms with their faces parallel to the brachy axis. A *primary* brachydome (~a : b : c) ({011}) is derived from the primary rhombic protopyramid (a : b : c).

- From the macropyramids (a : nb : mc) we can derive the corresponding macrodomes (a : ~b : mc) ({hol}) by making n infinite, which means that we get horizontal prisms with their faces parallel to the macro axis. A *primary* macrodome (a : ~b : c) is derived from the primary rhombic protopyramid (a : b : c).

The next Figures illustrate the derivation of the primary brachydome, and of the primary macrodome from the primary rhombic protopyramid.

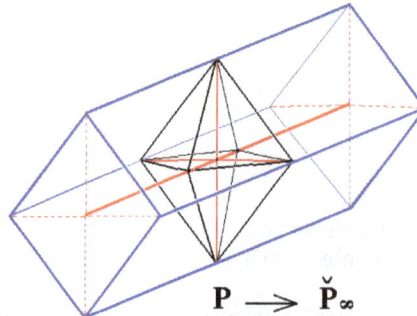

$$P \longrightarrow \check{P}_\infty$$

Derivation of the primary Brachydome (in which m = 1) from the primary Protopyramid. It is an open Form
(front side and back side open) -- the straight lines bordering its front and rear ends should not suggest a front
and back face -- consisting of four faces parallel to the brachy axis. It is a horizontal rhombic prism.

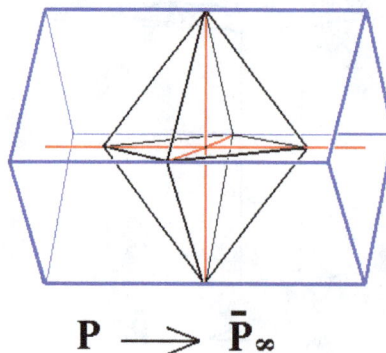

$$P \longrightarrow \bar{P}_\infty$$

Derivation of the primary Macrodome (in which m = 1) from the primary Protopyramid. It is an open Form
(left side and right side open) -- the straight lines bordering its left and right ends should not suggest a left
and right face -- consisting of four faces parallel to the macro axis. It is a horizontal rhombic prism.

Finally we can derive three more Forms of this Crystal Class, the pinacoids. They are Forms in which each face intersects only one and the same crystallogaphic axis :

- The Brachy Pinacoid.

- The Macro Pinacoid.

- The Basic Pinacoid.

From the brachydomes (~a : b : mc) (or from the primary brachydome (~a : b : c), for that matter), we can derive the brachy pinacoid (~a : b : ~c) ({010}), by letting m (which equals 1 in the primary brachydome) become infinite. It consists of two vertical faces parallel to each other and to the brachy axis.

To show the derivation, let us first depict the above constructed (primary) brachydome as it is all by itself. The Figure shows a shortened version of it which is immaterial because the length of the (horizontal) prism is not determined.

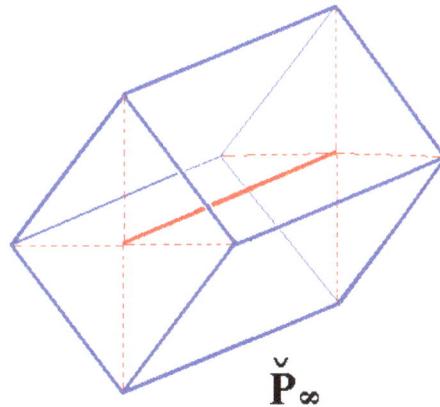

$$\check{\mathbf{P}}_\infty$$

The primary Brachydome. The red solid line is the brachy axis.

From this brachydome we can now derive the brachy pinacoid :

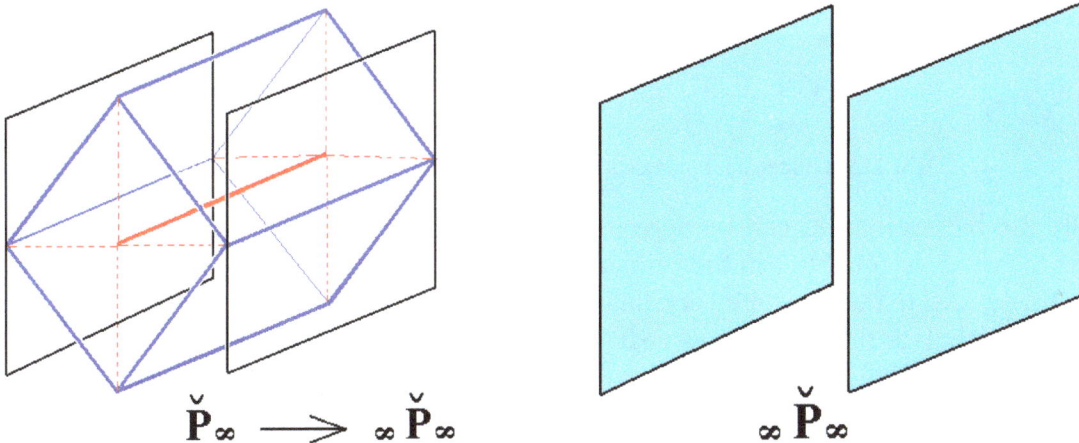

$$\check{\mathbf{P}}_\infty \longrightarrow {}_\infty\check{\mathbf{P}}_\infty \qquad\qquad {}_\infty\check{\mathbf{P}}_\infty$$

Derivation of the Brachy Pinacoid from the primary   The Brachy Pinacoid. This Form consists of two vertical
Brachydome. The red solid line is the brachy axis.   faces parallel to the brachy axis.

Like the other pinacoids and all the prisms, the brachy pinacoid is an open Form and can only exist in real crystals when combined with Forms of this Crystal Class such that the combination is a closed structure.

From the macrodomes (a :~b : mc) (or from the primary macrodome (a :~b : c), for that matter), we can derive the macro pinacoid (a :~b : ~c) ({100}), by letting m (which is equal to 1 in the primary macrodome) become infinite . It consists of two vertical faces parallel to each other and to the macro axis.

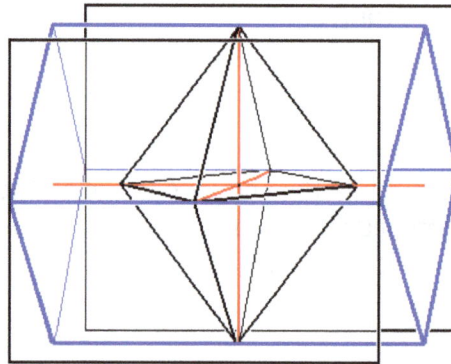

$$\overline{\mathbf{P}}_{\infty} \longrightarrow {}_{\infty}\overline{\mathbf{P}}_{\infty}$$

Derivation of the Macro Pinacoid from the primary Macrodome.
The red lines indicate the *crystallographic axes*.

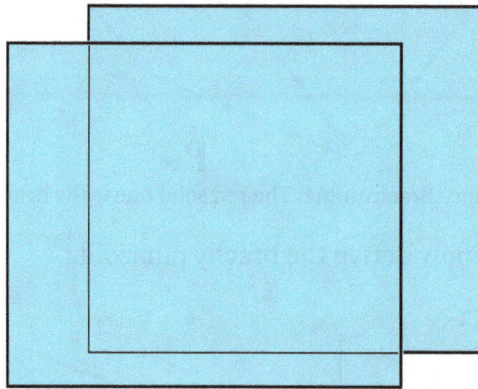

$${}_{\infty}\overline{\mathbf{P}}_{\infty}$$

The Macro Pinacoid. It consists of two vertical faces parallel to the macro axis.

Finally the basic pinacoid can be derived from either the brachydome or the macrodome by letting m (which is equal to 1 in the primary domes) become zero. The result will be a horizontal face pair, parallel to the horizontal crystallographic axes (the brachy axis and the macro axis). The Weissian symbol for this Form is (~a : ~b : c).

We have let the derivation coefficient m become zero, resulting in the faces ~a : ~b : +/-c to become parallel to the plane of the horizontal crystallographic axes. So in fact these faces should coincide, resulting in just one face with a zero coefficient of c. But because m = 0 only means that the face is horizontally oriented without its position being determined therewith, we can still see them as two faces -- horizontal faces -- and place them at unit distance above and below the origin of the system of crystallographic axes, causing the Weissian symbol of this Form to be (~a : ~b : c), and not (~a : ~b : 0c). Only then the resulting Form -- two horizontal (equivalent) faces -- complies with the symmetry content of the present Class.

The Naumann symbol is 0P, and the Miller symbol is {001}.

In the next Figure we will derive the basic pinacoid from the (primary) brachydome :

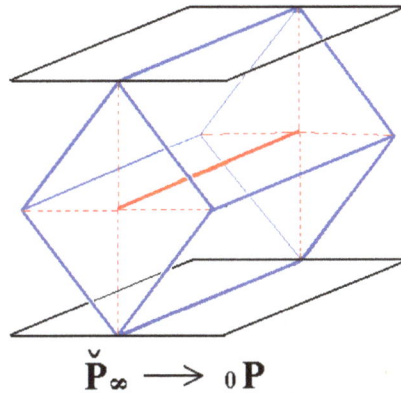

$$\check{P}_\infty \longrightarrow {}_0P$$

Derivation of the Basic Pinacoid from the primary Brachy Dome.

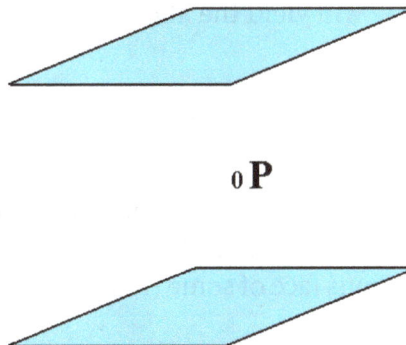

$${}_0P$$

The Basic Pinacoid. It consists of two horizontal faces parallel
to the brachy and macro axes.

We now have derived all the Forms of the Rhombic-bipyramidal Crystal Class (= Holohedric Division). These Forms all comply with the symmetry of this Class (2/m 2/m 2/m). And all these Forms can enter in combinations with each other in real Crystals.

The faces that represent each Form are -- with respect to the Weissian symbolism -- placed between brackets (as is done above), for example (~a : ~b : c). As such they are Forms.

These faces themselves -- thus without brackets, i.e. as faces -- are then the eleven Basic Faces compatible with the Orthorhombic Crystal System.

To sum up these basic faces we get :

1. a : b : mc

2. na : b : mc

3. a : nb : mc

4. $a : b : \sim c$

5. $na : b : \sim c$

6. $a : nb : \sim c$

7. $\sim a : b : mc$

8. $a : \sim b : mc$

9. $\sim a : b : \sim c$

10. $a : \sim b : \sim c$

11. $\sim a : \sim b : c$

They represent all possible configurations of derivation coefficients among the three crystallographic axes.

Subjecting each of these basic faces to the symmetry elements of the present Crystal Class -- which means generating new faces according to the symmetry demands of that Class imposed on the resulting face configuration (Form), i.e. the symmetry that this configuration should have according to those demands -- will yield the above Forms, i.e. the Forms of the Holohedric Division.

Subjecting those same basic faces to the symmetry elements of the other orthorhombic Crystal Classes will result in the Forms of those Classes.

The face a : b : c is the primary face (yielding the primary rhombic protopyramid, (a : b : c), when subjected to the symmetry elements of the present Class). This face, from which all listed faces are derivations, is taken from a conspicuous face of some (real) crystal belonging to the present Crystal Class.

That the above list of basic faces is complete can be shown by the location of their poles in the stereographic projection of the symmetry elements of the present Class :

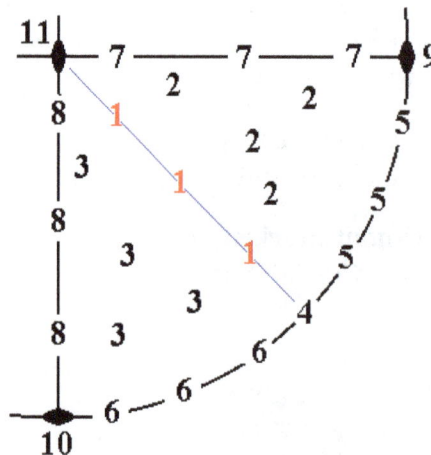

The lower right quadrant of the stereographic projection of the symmetry elements of the Rhombic-bipyramidal Class, and the possible locations of faces (face categories).

1. a : b : mc
2. na : b : mc
3. a : nb : mc
4. a : b : ~c
5. na : b : ~c
6. a : nb : ~c
7. ~a : b : mc
8. a : ~b : mc
9. ~a : b : ~c
10. a : ~b : ~c
11. ~a : ~b : c

In the above Figure the straight solid black lines are vertical mirror planes, perpendicular to each other. The solid circumference of the the circle (one quarter shown) represents the equatorial mirror plane. The black solid ellipses signify 2-fold rotation axes. Two of them are horizontal and coincide with the two horizontal crystallogaphic axes (the vertically drawn axis is the brachy axis, the horizontally drawn axis is the macro axis). The third 2-fold rotation axis is vertical and coincides with the c axis, it is perpendicular to the plane of the drawing.

1 represents the faces (face category) a : b : mc. They can vary along the line bisecting the quadrant in two equal halves. They all cut off unit distances of the brachy and macro axes, and can vary with respect to the cut-off distance of the c axis. All these distances are given in the form of a ratio of the three axial cut-off distances. If m = 1 then we have the primary basic face, that generates a primary rhombic protopyramid (a bipyramid) when subjected to the symmetry elements of the present Class. Other finite non-zero values of m will yield derived protopyramids.

2 represents the faces (face category) na : b : mc. Such a face can be everywhere inside the upper sector of the quadrant. When subjected to the symmetry elements of the present Class such a face will yield a brachy pyramid.

3 represents the faces (face category) a : nb : mc. Such a face can be everywhere in the lower sector of the quadrant. When subjected to the symmetry elements of the present Class such a face will yield a macro pyramid.

4 represents the face (face category) a : b : ~c. It is vertical and cuts off unit distances from the brachy and macro axes. Only one such face is possible. When it is subjected to the symmetry elements of the present Class it will yield a protoprism.

5 represents the faces (face category) na : b : ~c. They are vertical. When such a face is subjected to the symmetry elements of the present Class it will generate a brachy prism. It can vary along the circle segment bordering the upper sector of the quadrant.

6 represents the faces a : nb : ~c. They are also vertical. When such a face is subjected to the symmetry elements of the present Class it will generate a macro prism. It can vary along the circle segment bordering the lower sector of the quadrant.

7 represents the faces ~a : b : mc. They are parallel to the brachy axis. When such a face is subjected to the symmetry elements of the present Class it will generate a brachydome. It can vary along the the horizontal line in the drawing.

8 represents the faces a : ~b : mc. They are parallel to the macro axis. When such a face is subjected to the symmetry elements of the present Class it will generate a macrodome. It can vary along the the vertical line in the drawing.

9 represents the face ~a : b : ~c. It is vertical and parallel to the brachy axis. Only one such face is

possible. When this face is subjected to the symmetry elements of the present Class it will generate a brachy pinacoid.

10 represents the face a : ~b : ~c. It is vertical and parallel to the macro axis. Only one such face is possible. When this face is subjected to the symmetry elements of the present Class it will generate a macro pinacoid.

11 represents the face ~a : ~b : c. It is horizontal. Only one such face is possible. When this face is subjected to the symmetry elements of the present Class it will generate a basic pinacoid.

Remark about the Stereographic Projection of Orthorhombic Crystals

In contrast with the Isometric, Tetragonal and Hexagonal Systems, all the crystallographic axes of the Orthorhombic System are non-equivalent. This means that the absolute magnitude of the unit intersection distances are not necessarily the same for all three axes. So when we have a face a : b : c (= (111)) of an orthorhombic crystal, then the absolute cut-off distances with respect to the three axes are not necessarily the same although the equality of the derivation coefficients (and also of the corresponding Miller indices) might suggest so.

Figure shows the stereographic projection of some faces of an orthorhombic crystal where indeed we have to do with unequal cut-off distances connected with equal derivation coefficients and Miller indices.

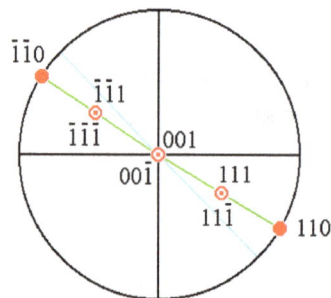

Stereographic projection of the faces a : b : c (= (111)), a : b : -c (= (111*)) (where 1* means a negative Miller index, written as a 1 with a score above it in the literature),   -a : -b : c(= (1*1*1)), -a : -b : -c (= (1*1*1*)), a : b : ~c (= (110)) and   -a : -b : ~c (= (1*1*0))   of an orthorhombic crystal.

These faces all have equal derivation coefficients, respectively Miller indices, with a value equal to 1, negative or positive. The absolute cut-off distances are, however, not the same in this case as is to be expected for orthorhombic crystals. So those faces do not lie on the bisector (given in blue) of the relevant quadrants of the projection plane. They do lie, however, on one straight line (given in green), a diameter of the projection plane.

Also the projections of the two horizontal faces ~a : ~b : c (= (001)) and ~a : ~b : -c (= (001)) are* shown.

In the stereographic projections that follow we will however place the faces with equal derivation coefficients with respect to the brachy axis (= a axis) and macro axis (= b axis) on the bisector of the relevant quadrant of the projection plane (Real crystals actually showing such positions of the relevant faces are not principally excluded in the Orthorhombic System). This we do to let the equality of the derivation coefficients stand out clearly.

Let us now actually execute the above mentioned generations of the Forms when each basic face is subjected to the symmetry elements of the present Class. Of course we'll end up with the same Forms that we have already derived above. But deriving those Forms from the basic faces better shows the fact that the set of derived Forms is complete.

The derivation will be done by means of stereographic projections of those basic faces (in fact their face poles) and of the symmetry elements of our Class. As should be clear we will thus follow the facial approach in deriving Forms as we did in the Tetragonal and Hexagonal Crystal Systems.

Facial Approach

The face a : b : mc generates a (derived) protopyramid when subjected to the symmetry elements of the present Class. First the face is duplicated by one vertical mirror plane, then the two faces are again duplicated by the other vertical mirror plane, resulting in four faces, forming a monopyramid. Then this monopyramid is reflected in the horizontal mirror plane resulting in a bipyramid. The rest of the symmetry elements is then implied.

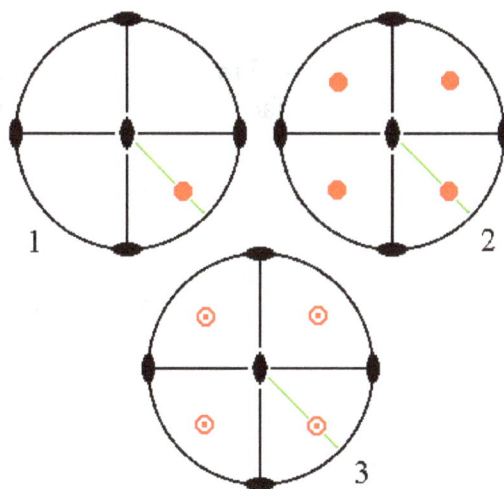

(1). Position of the face a : b : mc in the stereogram of the symmetry elements of the Rhombic-bi-pyramidal Crystal Class.

(2). Generation of four faces in virtue of the action of the two vertical mirror planes (straight solid black lines).

(3). Reflection of those four faces in the equatorial mirror plane (black solid circle).

The small centred red circles indicate a pair of upper and lower faces.

The face na : b : mc generates a brachy pyramid when subjected to the symmetry elements of the present Class. The face is duplicated by one vertical mirror plane. Then the resulted face pair is duplicated by the other vertical mirror plane. Finally the four resulted faces are reflected in the equatorial mirror plane resulting in a bipyramid. The other symmetry elements are then implied.

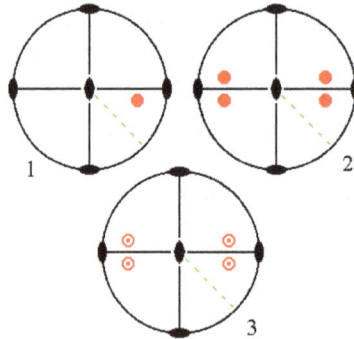

(1). Position of the face na : b : mc in the stereogram of the symmetry elements of the Rhombic-bi-pyramidal Crystal Class.

(2). Generation of four faces in virtue of the action of the two vertical mirror planes (straight solid black lines).

(3). Reflection of those four faces in the equatorial mirror plane (black solid circle).

The face a : nb : mc generates a macropyramid when subjected to the symmetry elements of the present Class. It will be duplicated by one vertical mirror plane and again duplicated by the other vertical mirror plane. Then the resulted four faces are reflected in the equatorial mirror plane resulting in a bipyramid. The other symmetry elements are then implied.

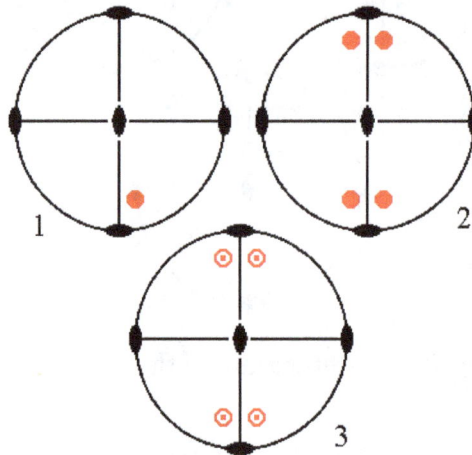

(1). Position of the face a : nb : mc in the stereogram of the symmetry elements of the Rhombic-bi-pyramidal Crystal Class.

(2). Generation of four faces in virtue of the action of the two vertical mirror planes (straight solid black lines).

(3). Reflection of those four faces in the equatorial mirror plane (black solid circle).

The face a : b : ~c is vertical. It generates the rhombic protoprism when subjected to the symmetry elements of the present Class : The face is duplicated by one vertical mirror plane, and the resulted face pair is again duplicated (now) by the other vertical mirror plane, resulting in four vertical faces. The other symmetry elements are then implied.

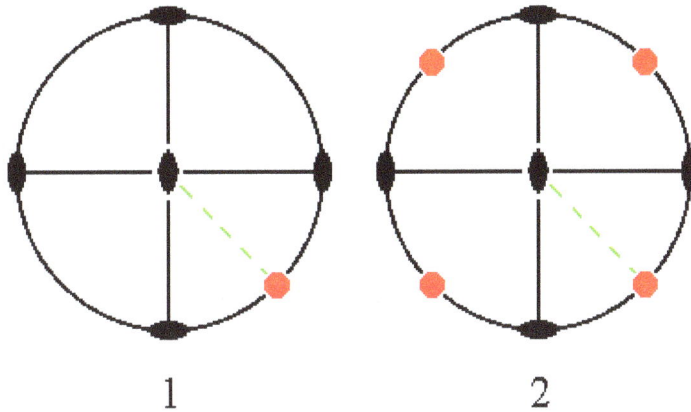

1   2

(1). Position of the face a : b : ~c in the stereogram of the symmetry elements of the Rhombic-bi-pyramidal Crystal Class.

(2). Generation of four faces in virtue of the action of the two vertical mirror planes, making up a prism.

The face na : b : ~c is also vertical. It generates a brachyprism when subjected to the symmetry elements of the present Class : The face is duplicated by one vertical mirror plane, and the resulted face pair is again duplicated (now) by the other vertical mirror plane, resulting in four vertical faces making up a prism.

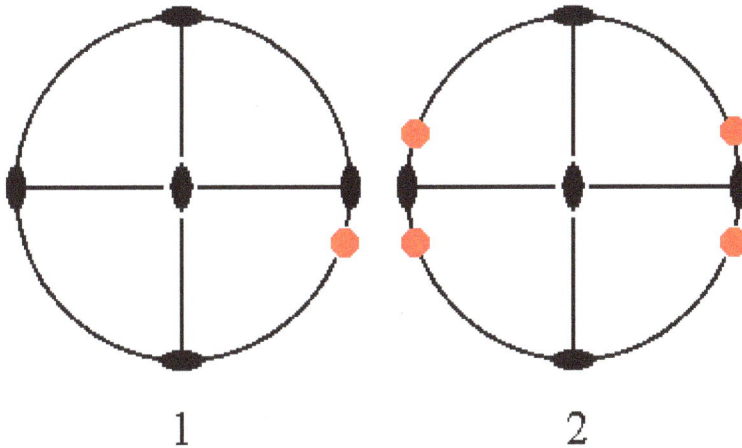

1   2

(1). Position of the face na : b : ~c in the stereogram of the symmetry elements of the Rhombic-bi-pyramidal Crystal Class.

(2). Generation of four faces in virtue of the action of the two vertical mirror planes, making up a prism.

The face a : nb : ~c is also vertical. It generates a macroprism when subjected to the symmetry elements of the present Class: The face will be duplicated by one vertical mirror plane and the result is again duplicated (now) by the other vertical mirror plane, resulting in four vertical faces making up a prism. All other symmetries of the Class are then implied.

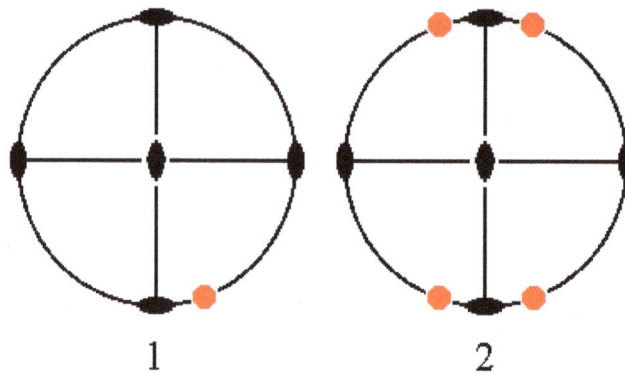

1   2

(1). Position of the face a : nb : ~c in the stereogram of the symmetry elements of the Rhombic-bi-pyramidal Crystal Class.

(2). Generation of four faces in virtue of the action of the two vertical mirror planes, making up a prism.

The face ~a : b : mc is parallel to the brachy axis. It generates a brachydome when subjected to the symmetry elements of the present Class : The face is duplicated by the vertical mirror plane in which the brachy axis lies (or, having the same effect in this case, it is duplicated by the vertical 2-fold rotation axis). The resulting face pair is then reflected in the equatorial mirror plane, yielding four faces parallel to the brachy axis making up a horizontal prism, in this case a brachydome.

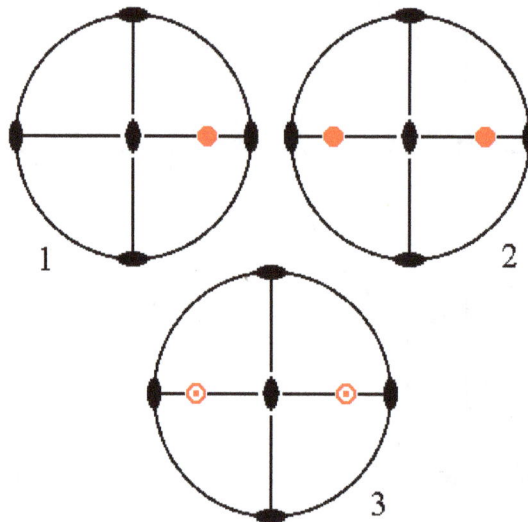

(1). Position of the face ~a : b : mc in the stereogram of the symmetry elements of the Rhombic-bi-pyramidal Crystal Class.

(2). Duplication of the face in virtue of the action of the vertical mirror plane in which the brachy axis lies.

(3). The resulted face pair is then in turn duplicated by the action of the equatorial mirror plane yielding four faces parallel to the brachy axis, making up a horizontal prism.

The face a : ~b : mc is parallel to the macro axis. It generates a macrodome when subjected to the symmetry elements of the present Class : The face is duplicated by the action of the vertical mirror plane in which the macro axis lies (or, giving the same result, is duplicated by the action of the vertical 2-fold rotation axis). Then the resulted face pair is reflected in the equatorial mirror plane yielding four faces parallel to the macro axis, a horizontal prism, in the present case a macrodome.

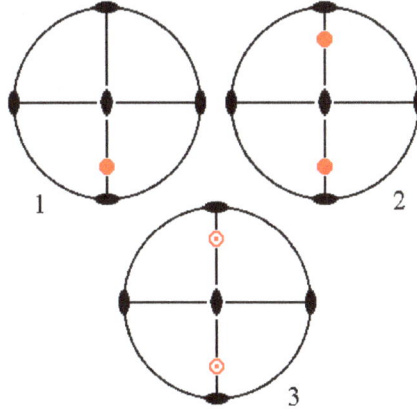

(1). Position of the face a : ~b : mc in the stereogram of the symmetry elements of the Rhombic-bi-pyramidal Crystal Class.

(2). Duplication of the face in virtue of the action of the vertical mirror plane in which the macro axis lies.

(3). The resulted face pair is then in turn duplicated by the action of the equatorial mirror plane yielding four faces parallel to the macro axis, making up a horizontal prism.

The face ~a : b : ~c is vertical and parallel to the brachy axis. It will generate a brachy pina-coid when subjected to the symmetry elements of the present Class :   The face is duplicated by the vertical mirror plane that contains the brachy axis (or, having the same effect, it is duplicated by the vertical 2-fold rotation axis). All other symmetries are now implied. The result is a Form consisting of two vertical faces parallel to the brachy axis, the brachy pinacoid.

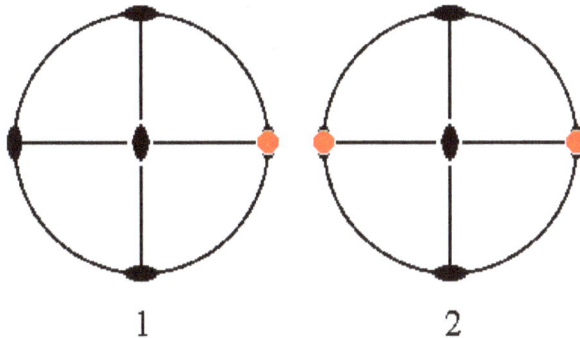

(1). Position of the face ~a : b : ~c in the stereogram of the symmetry elements of the Rhombic-bi-pyramidal Crystal Class.

(2). Duplication of the face in virtue of the action of the vertical mirror plane in which the brachy axis lies, resulting in a vertical face pair parallel to the brachy axis.

The face a : ~b : ~c is vertical and parallel to the macro axis. It generates a macropinacoid when subjected to the symmetry elements of the present class : The face is duplicated by the action of the vertical mirror plane that contains the macro axis (or, giving the same effect, it is duplicated by the action of the vertical 2-fold rotation axis), resulting in a vertical face pair parallel to the macro axis, making up the macro pinacoid.

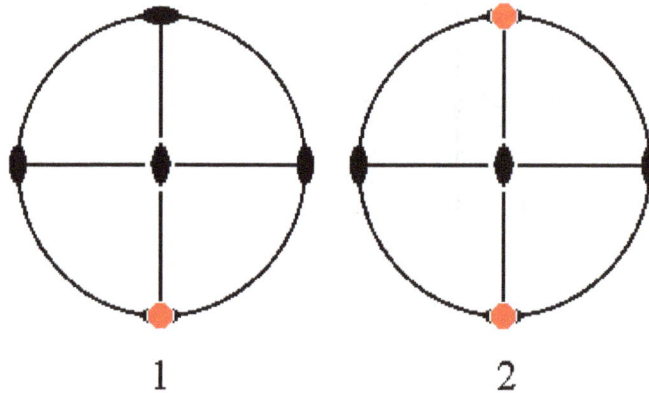

1   2

(1). Position of the face a : ~b : ~c in the stereogram of the symmetry elements of the Rhombic-bi-pyramidal Crystal Class.

(2). Duplication of the face in virtue of the action of the vertical mirror plane in which the macro axis lies, resulting in a vertical face pair parallel to the macro axis.

The face ~a : ~b : c, finally, is horizontal. It generates a basic pinacoid when subjected to the symmetry elements of the present Class : The face is reflected in the equatorial mirror plane, resulting in a horizontal face pair, the basic pinacoid. The other symmetries of the Class are then implied.

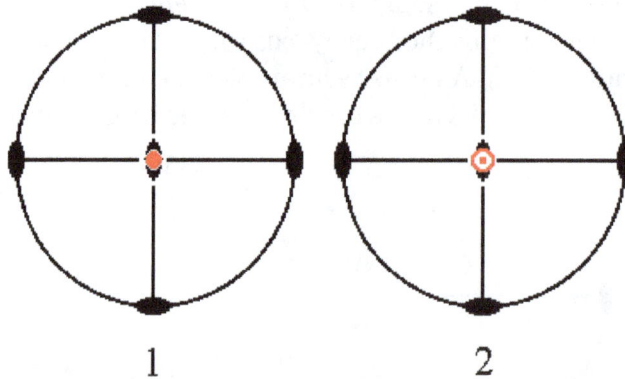

1   2

(1). Position of the face ~a : ~b : c in the stereogram of the symmetry elements of the Rhombic-bi-pyramidal Crystal Class.

(2). Duplication of the face in virtue of the action of the horizontal mirror plane (that contains the brachy and the macro axis) resulting in a horizontal face pair parallel to the brachy and macro axis.

This concludes our exposition of the Rhombic-bipyramidal Crystal Class (= Holohedric Division of the Orthorhombic Crystal System).

Monoclinic Crystal System

Monoclinic system, one of the structural categories to which crystalline solids can be assigned. Crystals in this system are referred to three axes of unequal lengths—say, a, b, and c—of which a is perpendicular to b and c, but b and c are not perpendicular to each other.

If the atoms or atom groups in the solid are represented by points and the points are connected, the resulting lattice will consist of an orderly stacking of blocks, or unit cells. The monoclinic unit cell is distinguished by a single axis, called an axis of twofold symmetry, about which the cell can be rotated by 180° without changing its appearance. More solids belong to the monoclinic system than to any other. Beta-sulfur, gypsum, borax, orthoclase, kaolin, muscovite, clinoamphibole, clinopyroxene, jadeite, azurite, and spodumene crystallize in the monoclinic system.

Monoclinic crystal system

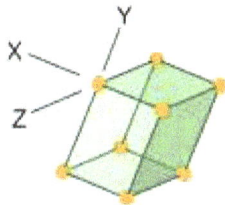

All three axes are unequal in length, and two axes are perpendicular to each other.

Its one variant is:

Base-centred monoclinic
lattice points in the middle of each of the two ends

Monoclinic crystal system; Crystals in a monoclinic system are referred to three axes of unequal lengths, with two axes being perpendicular to each other.

The Basic Form of the Monoclinic Crystal System. From this Form all other Forms of the Holohedric Division will be derived by changing the derivation coefficients. We can also consider just the one or the other hemipyramid as the Basic Form and derive all other Forms from it.

All these Forms can be represented by one of their faces (we choose the one with the least number of minus signs) and in this way we will obtain a list of Basic Faces compatible with the Monoclinic Crystal System. As a check we then will derive the same Forms (of the Holohedric Division) by subjecting these basic faces one by one to the symmetry elements of the Holohedric Division (= Monoclinic-prismatic Class).

The Forms of the two lower symmetrical Classes of our System will be derived by applying respectively hemimorphy and hemihedric to the holohedric Forms (merohedric approach) and (as a check) again by subjecting the just mentioned basis faces one by one to the symmetry elements of those (lower symmetrical) Classes.

In order to precisely understand the face configuration of the Basic Form, the Monoclinic Pyramid, (and of the configuration of the Forms that combine to give this Basic Form) we will give a drawing of them, as is done on Figure.

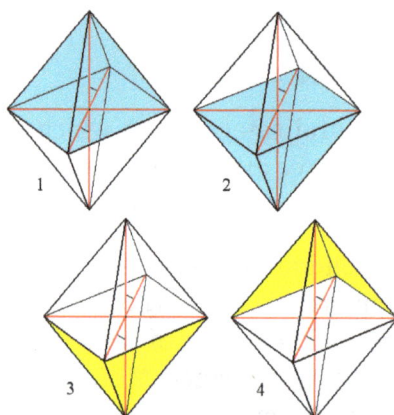

The Monoclinic Pyramid (four times depicted).

It is a bipyramid consisting of eight faces each of them intersecting with all three crystallographic axes. It is in fact a combination of two Forms, each consisting of four faces. One such Form is indicated in blue, the other in yellow.

The monoclinic pyramid is, as has been said, the Basic Form of the Monoclinic Crystal System. It is a bipyramid consisting of eight faces. It is that Form the faces of which intersect all three crystallogaphic axes. It differs from the pyramids of other Systems, for example the orthorhombic pyramid, first of all by the fact that its middle edges do not lie in a plane that is perpendicular to the vertical axis, but, corresponding with the inclination of the clino axis, in an oblique plane, and, because the ortho axis is perpendicular to the clino axis its outline is a rhombus. But the greatest difference lies in the already mentioned fact that the monoclinic pyramid is not a simple Form but a combination of two Forms. This is because the symmetry of the highest symmetrical Class (= Holohedric Division, 2/m) demands that to one face a : b : c belongs just one symmetrically positioned second face, and to this face pair belongs in turn a second face pair because of the 2-fold rotation axis (or, equivalently, because of the center of symmetry). So the resulting Form consists of just four faces (instead of eight in the case of the bipyramids of some other Systems). The other four faces of the monoclinic pyramid represent a second Form, derived from the face a : b : -c. The generation of the first (mentioned) Form, i.e. the hemipyramid derived from the face a : b : c is depicted in Figure. Its faces lie in the obtuse angle beta (obtaining between the vertical axis and the clino axis) and, to express this, the Form is indicated by a "-" sign preceding P in the Naumann symbol.

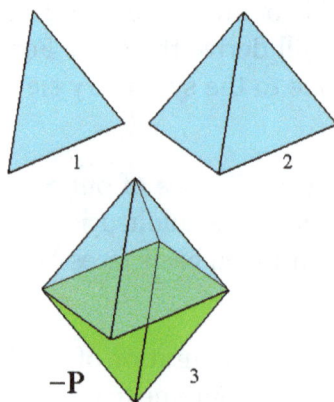

(1). The face a : b : c. (2). Duplication of that face by reflection in the mirror plane (that is horizontal), resulting in a face pair. (3). Duplication of this face pair by the action of the center of symmetry (which demands that every face has its parallel counter face). The same effect is obtained by the action of the 2-fold rotation axis (that is horizontal, and perpendicular to the mirror plane) by rotation about it of 1800, resulting in a negative Monoclinic Hemipyramid.

The generation of the other hemipyramid is depicted in Figure, and the final result once again in next Figure (the initial face differs from the one above in its having -c (minus c) instead of just c). The faces of this Form (this second hemipyramid) lie in the acute angle beta (obtaining between the vertical axis and the clino axis), and to express this (difference with the first hemipyramid) the Form is indicated by a "+" sign preceding the P in the Naumann symbol :

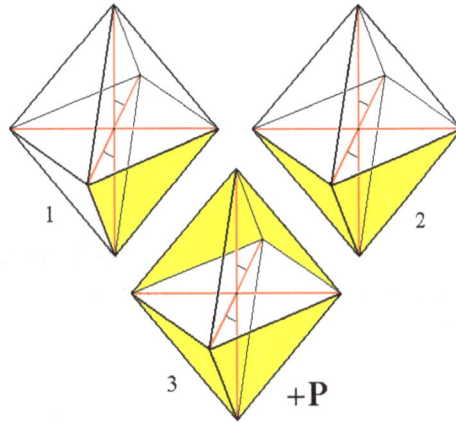

(1). The face a : b : -c.

(2). Duplication of that face by reflection in the mirror plane (that is horizontal), resulting in a face pair.

(3). Duplication of this face pair by the action of the center of symmetry (which demands that every face has its parallel counter face). The same effect is obtained by the action of the 2-fold rotation axis (that is horizontal, and perpendicular to the mirror plane) by rotation about it of 1800, resulting in a positive Monoclinic Hemipyramid.

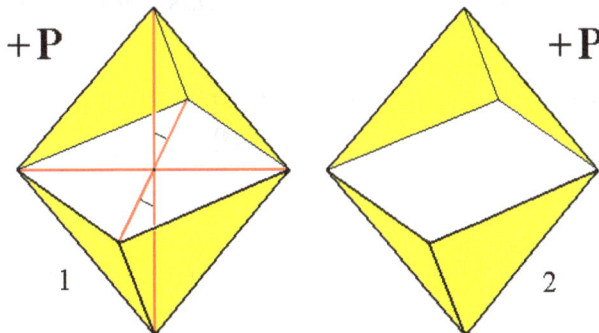

The final result of the generation of the positive Monoclinic Hemipyramid.

(1). Crystallographic axes shown. (2). Crystallographic axes not shown.

So the monoclinic pyramid (having eight faces) decays into two independent hemipyramids of which one consists of the following faces lying in the obtuse angle beta :

- a : b : c (upper right front face)

- a : -b : c (upper left front face)

- -a : -b : -c (lower left back face)

- -a : b : -c (lower right back face)

It is indicated with the Weissian symbol (a : b : c), with the Naumann symbol -P, and with the Miller symbol {111}.

The other hemipyramid consists of the following faces lying in the acute angle beta :

- -a : -b : c (upper left back face)

- -a : b : c (upper right back face)

- a : b : -c (lower right front face)

- a : -b : -c (lower left front face)

It is indicated with the Weissian symbol (a : b : -c), with the Naumann symbol +P, and with the Miller symbol {111*} (in which the sign "*" means a minus sign, indicated as a score above the relevant numeral in the literature.

The Monoclinic-prismatic Class (= Holohedric Division) 2/m

The symmetry elements of this Class are the folowing :

- One mirror plane.

- One 2-fold rotation axis, perpendicular to the mirror plane, and coincident with the crystallographic ortho axis.

- Center of symmetry.

From the above discussed Forms, -P and +P we can derive -- like we did in the Rhombic System -- three Series of hemipyramids (the above discussed Forms belong -- as primary ones) to the first one of these Series) :

- Hemipyramids of the Vertical Series $\pm mP$ (i.e. +mP and -mP) (protohemipyramids).

- Hemipyramids of the Clinodiagonal Series $\pm m\mathring{P}n$ (clinohemipyramids.

- Hemipyramids of the Orthodiagonal Series $\pm m\mathring{P}n$ (orthohemipyramids).

When the sigh "n" occurs in a Naumann symbol then it should, like in the Orthorhombic System, be greater than 1 (this can be accomplished by multiplying all three derivation coefficients with an appropriate number). When no such sign is present like in +mP then the relevant derivation coefficient is equal to 1.

In the Vertical Series the derivation coefficient m (referring to the vertical axis) is varied, while the (relative) cut-off distances (measured from the origin of the axial system) along the clino axis as well as along the ortho axis are of unit length (i.e. their derivation coefficients are equal to 1). This

Series can be represented by the face a : b : mc which is the first face we list here as a basic face (The list is given below).

To depict a member of the Vertical Series, the protohemipyramids, we repeat the above Figure of a primary protohemipyramid (in which m = 1), namely the positive one (+P) :

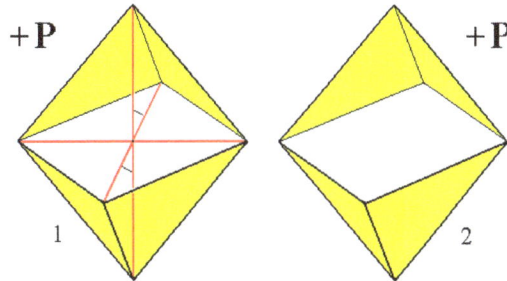

The positive primary Monoclinic Protohemipyramid. (1). Crystallographic axes shown. (2). Crystallographic axes not shown. The Form consists of four faces and is an open Form. Also a negative Monoclinic Protohemipyramid exists.

The Weissian symbol of this positive primary protohemipyramid is (a : b : -c), the Naumann symbol is +P, and the Miller symbol is {111*}. The *generalized* monoclinic protohemipyramid, i.e. a derived protohemipyramid, is depicted. This one will figure as the first listed Form (this list will be given below) of the Monoclinic-prismatic Crystal Class, and it can represent the Vertical Series of hemipyramids. Its Weissian symbol is (a : b : -mc) and its Naumann symbol is given in the Figure. Of course also the negative protohemipyramid, (a : b : mc), can represent the Vertical Series.

The Forms just discussed were considered to be *pyramids* (hemipyramids) because not any face of them is parallel to some crystallographic axis. But the shape of any monoclinic hemipyramid is equivalent to that of a prism. But such a prism is not a vertical prism but a tilted prism as the next Figure illustrates :

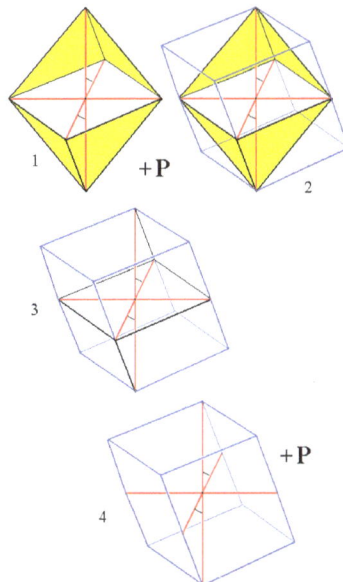

(1). The (primary) Protohemipyramid. (2). Construction of the equivalent prism. (3). Construction almost finished. (4). Final result of the construction. Red lines represent the crystallographic axes.

In the same way the next to be discussed hemipyramids of the Clinodiagonal and Orthodiagonal Series can be equally considered as (tilted) prisms. The most general Form of the present Crystal Class is either a hemipyramid of the Clinodiagonal Series or of the Orthodiagonal Series. And because these hemipyramids are, with respect to shape, equivalent to prisms the present Class -- a Class is always named after its most general Form -- is called the Monoclinic-*prismatic* Class.

The Clinodiagonal Series of hemipyramids is derived from the Vertical Series by letting the derivation coefficient that refers to the clino axis vary, while the coefficient referring to the ortho axis is unity. This Series can be represented by the face $na : b : mc$, which is a next basic face (i.e. a next basic face in our successive listing of such faces). The next series of images depicts the stages of the construction (derivation) of a positive primary (where $m = 1$) clinohemipyramid from the positive primary ($m = 1$) protohemipyramid.

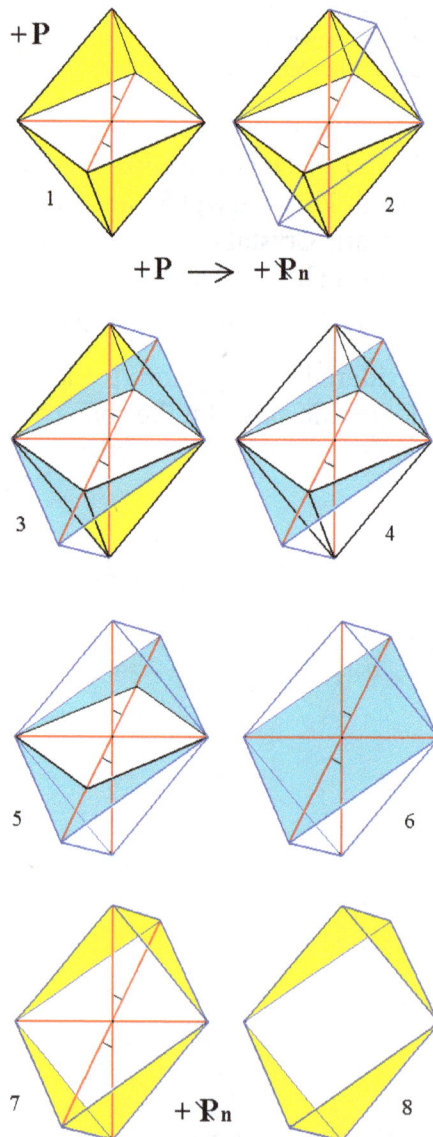

Stages in the construction of a positive primary Clinohemipyramid
from the positive primary protohemipyramid.

(1). The positive primary Protohemipyramid.

(2). Extention of the plane of the clino and ortho axes in the direction of the clino axis.

(3). Emphasis of enlarged clino-ortho plane.

(4). De-emphasizing the faces of the initial Form.

(5). Removal of the polar edges of the initial Form.

(6). Emphasis of the new clino-ortho plane.

(7). Emphasis of the faces of the new Form.

(8). Removal of the crystallographic axes. The result is a new Form, the positive primary Clinohemipyramid.

In the above Figure we derived a clinohemipyramid from a primary protohemipyramid, in which the coefficient m referring to the vertical axis is equal to 1. But in fact we can derive a clinohemipyramid from any protohemipyramid, which means that from any member of the Vertical Series (having m not equal to 1) a corresponding clinohemipyramid can be derived. In the next Figure we will first derive a non-primary protohemipyramid, and in the Figure thereafter we will derive a non-primary clinohemipyramid from it.

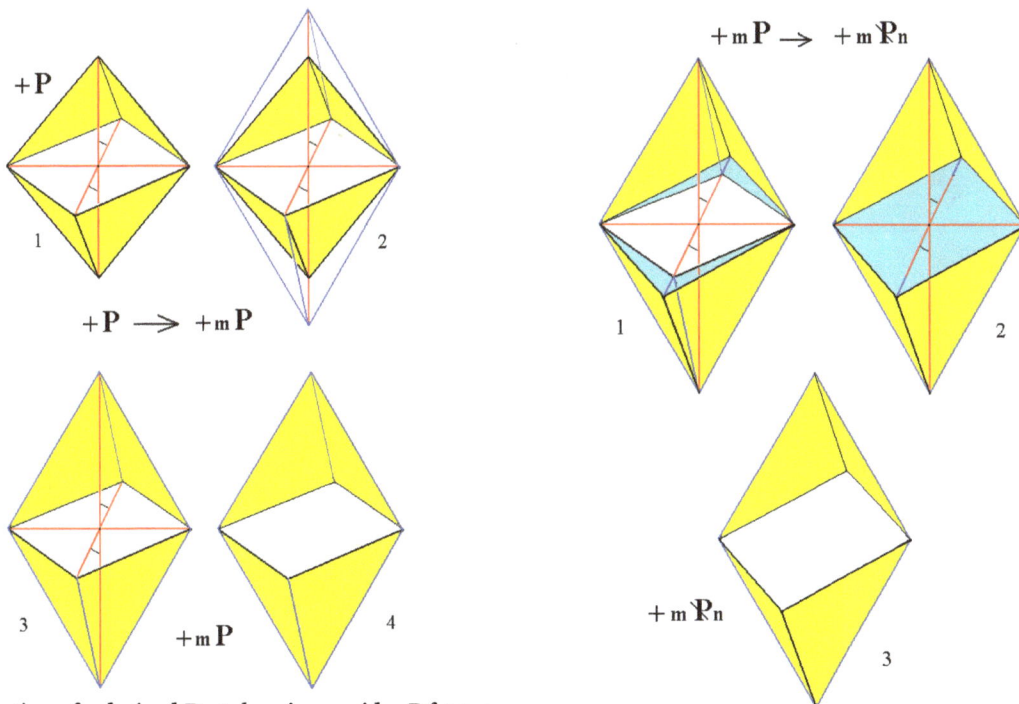

Construction of a derived Protohemipyramid mP from a primary Protohemipyramid P. The Crystallographic axes are indicated in red.

Construction of a derived (positive) Clinohemipyramid from a derived Protohemipyramid.

This derived (positive) clinohemipyramid will be the next Form of our listing of the Forms of the present Crystal Class. It can represent the Clinodiagonal Series of monoclinic hemipyramids. Its

Weissian symbol is (na : b : -mc), the Naumann symbol is given in the Figure, and the Miller symbol is {khl*} (for the negative hemipyramid it is {khl}). Of course also the negative derived clino-hemipyramid, (na : b : mc), can represent the Clinodiagonal Series.

The Orthodiagonal Series of hemipyramids is derived from the Vertical Series by letting the derivation coefficient that refers to the ortho axis vary, while the coefficient referring to the clino axis is unity. This Series can be represented by the face a : nb : mc, which is yet another basic face (i.e. a next basic face in our successive listing of such faces). The next series of images depicts the stages of the construction (derivation) of a positive non-primary (where m is unequal to 1) orthohemipyramid from the positive derived (m is unequal to 1) protohemipyramid.

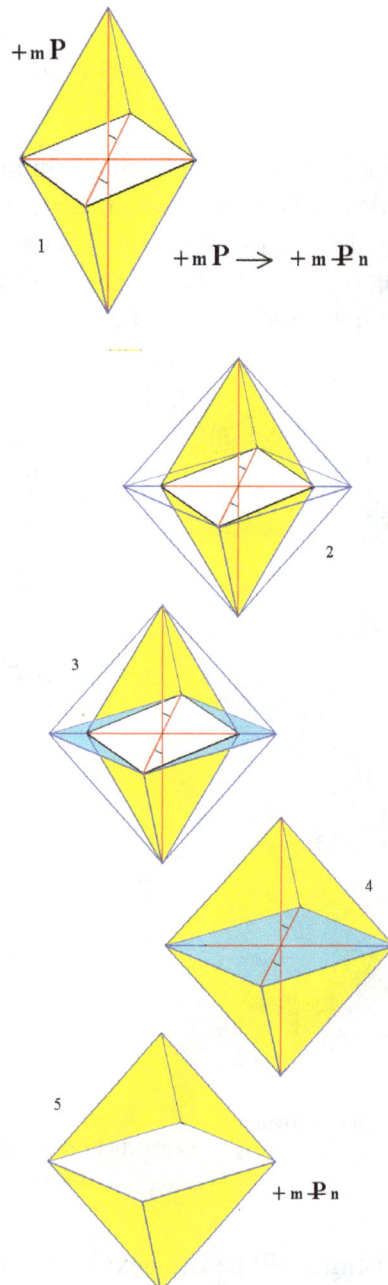

(1). The derived (i.e. m is unequal to 1) positive Protohemipyramid.

(2). Extension of the clino-ortho plane in the direction of the ortho axis.

(3). Emphasis of the extension.

(4). Emphasis of the new clino-ortho plane and the new faces.

(5). Crystallographic axes removed. The new Form -- the positive non-primary Orthohemipyramid -- consisting of four faces is clearly displayed.

This derived (positive) orthohemipyramid will be yet another Form of our listing of the Forms of the present Crystal Class. It can represent the Orthodiagonal Series of monoclinic hemipyramids. Its Weissian symbol is (a : nb : -mc), the Naumann symbol is given in the Figure, and the Miller symbol is {hkl*} (for the negative hemipyramid it is {hkl}). Of course also the negative derived orthohemipyramid, (a : nb : mc), can represent the Orthodiagonal Series.

Like the monoclinic hemipyramids, also the monoclinic vertical prisms come in three Series :

- Vertical Prism of the Vertical Series. This Series consists of only one member, the mono-clinic protoprism. It can be derived from any member of the Vertical Series of hemipyra-mids. Its Naumann symbol is ∞P .

- Vertical Prisms of the Clinodiagonal Series, the monoclinic clinoprisms, ∞P_n . They can be derived from the hemipyramids of the Clinodiagonal Series.

- Vertical Prisms of the Orthodiagonal Series, the monoclinic orthoprisms, $\infty \bar{P}_n$. They can be derived from the hemipyramids of the Orthodiagonal Series.

From the monoclinic protohemipyramid (as such belonging to the Vertical Series of hemipyramids) can be derived the monoclinic protoprism by letting the derivation coefficient m become infinite. The faces then become vertical, resulting in a prism with its four faces parallel to the vertical axis (c axis). Every protohemipyramid (positive or negative, primary or non-primary) leads to the same monoclinic protoprism, so there is only one such prism. Its Weissian symbol is (a : b : ~c) (where the sign"~" stands for infinity), its Naumann symbol is ~P, and the Miller symbol is {110}. With it we have found yet another basic face, namely   a : b : ~c, for our list of all the basic faces compatible with the Monoclinic Crystal System. The monoclinic protoprism is, like the monoclinic hemipyramids, an open Form consisting of four faces. The straight lines indicating the upper and lower borders of the prism should not suggest upper and lower faces (closing the prism). If such faces are present then they are a separate Form (the basic pinacoid, consisting of two faces) and combine with the prism resulting in a closed Form. With the monoclinic pro-toprism we have yet another Form in our listing of monoclinic holohedric Forms (i.e. Forms of the present Crystal Class). In the next Figure we construct the monoclinic protoprism from the derived positive protohemipyramid.

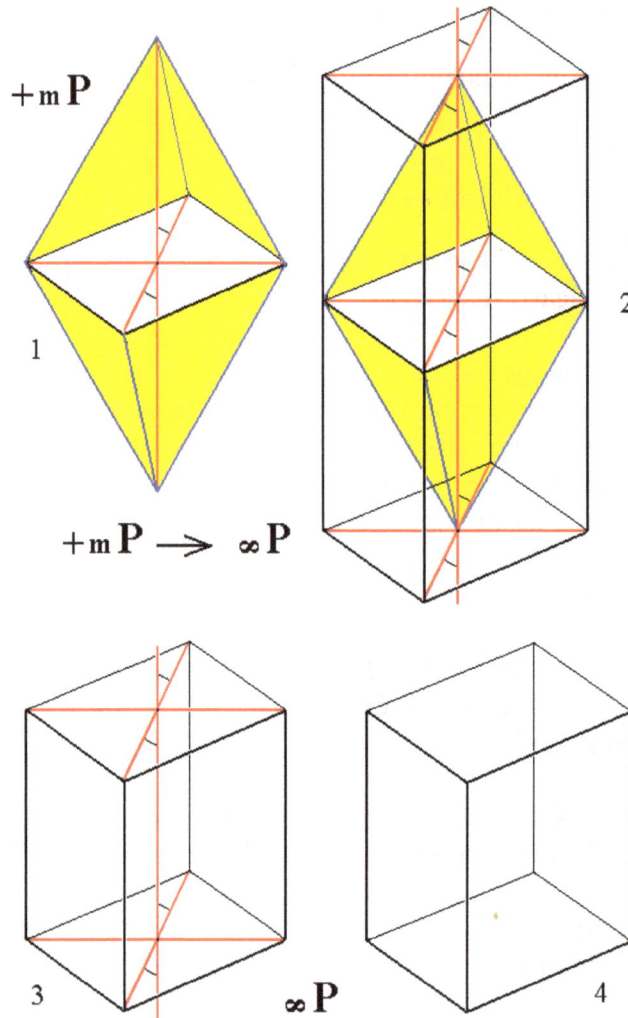

$+_m P$

$+_m P \twoheadrightarrow \infty P$

∞P

(1). The positive derived (i.e. m is not equal to 1) Monoclinic Protohemipyramid.

(2). Construction of the Monoclinic Protoprism from the Monoclinic Protohemipyramid.

(3). Result of the construction. The prism is drawn shortened along the vertical axis by reason of convenience (which does not make any difference crystallographically).

(4). The Monoclinic Protoprism. Axial system removed from the drawing.

From a clinohemipyramid we can derive a monoclinic clinoprism by letting the derivation coefficient m become infinite. The resulting prism has faces each of which is parallel to the vertical axis, cuts off a unit piece (positive or negative) from the ortho axis, and (cuts off) a piece from the clino axis that is longer than the unit piece associated with that axis. The Weissian symbol for these Forms is (na : b : ~c), the Naumann symbol is indicated in (2) of the above list of vertical prisms, and the Miller symbol is {kh0}. With all this we have found yet another basic face for our listing of such faces, namely   na : b : ~c. With the monoclinic clinoprism we have yet another Form for our listing of monoclinic holohedric Forms. In the next Figure we will construct a monoclinic clinoprism from the derived positive clinohemipyramid of Figure:

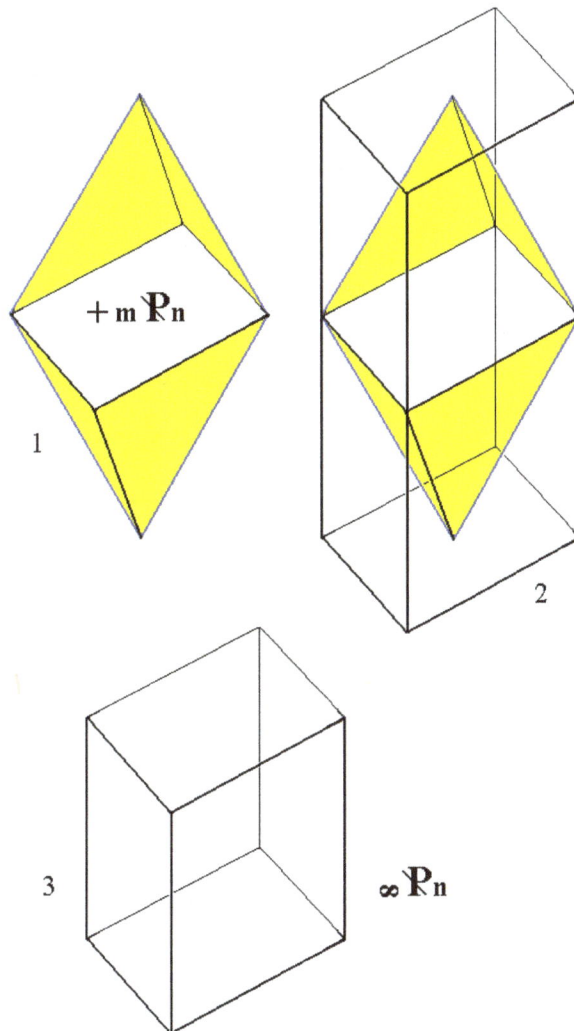

(1). The positive derived Clinohemipyramid.

(2). Construction of the Monoclinic Clinoprism from the positive derived Clinohemipyramid (The corresponding negative hemipyramid will give the same result).

(3). The resulting Monoclinic Clinoprism. The prism is drawn shortened along the vertical axis by reason of convenience (which does not make any difference crystallographically).

From an orthohemipyramid we can derive a monoclinic orthoprism by letting the derivation coefficient m become infinite. The resulting prism has faces each of which is parallel to the vertical axis, cuts off a unit piece (positive or negative) from the clino axis, and (cuts off) a piece from the ortho axis that is longer than the unit piece associated with that axis. The Weissian symbol for these Forms is (a : nb : ~c), the Naumann symbol is indicated in (3) of the above list of vertical prisms, and the Miller symbol is {hko}. With all this we have found yet another basic face, namely (a : nb : ~c). With the monoclinic orthoprism we have found yet another Form for our listing of holohedric monoclinic Forms. In the next Figure we will construct a monoclinic orthoprism from the derived positive orthohemipyramid of Figure:

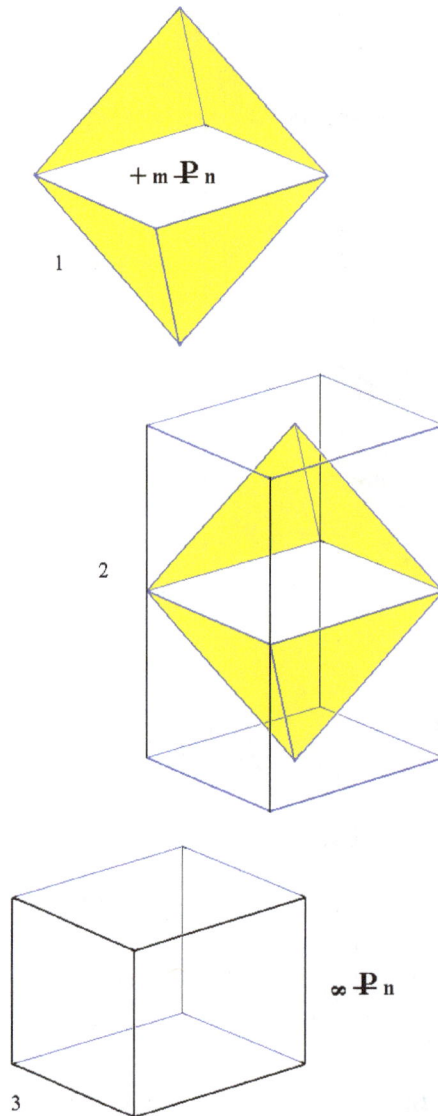

(1). The derived positive Orthohemipyramid.

(2). Construction of a Monoclinic Orthoprism from the derived positive Orthohemipyramid (The corresponding negative hemipyramid will give the same result).

(3). Result of the construction.

From the protohemipyramid we can also derive a prism that is not vertical but inclined according to the angle between the vertical and clino axes. It can be obtained by letting the derivation coefficient n, when it refers to the clino axis, become infinite. The result is a clinodome, i.e. a Form consisting of four faces parallel to the clino axis. Its Weissian symbol is (~a : b : mc), its Naumann symbol is $m\overset{\cdot}{P}\infty$, and the Miller symbol is {ohl}. With it we have found yet another basic face, namely ~a : b : mc, for our list of basic faces compatible with the Monoclinic Crystal System. With the monoclinic clinodome we have found yet another Form for our list of holohedric monoclinic Forms.

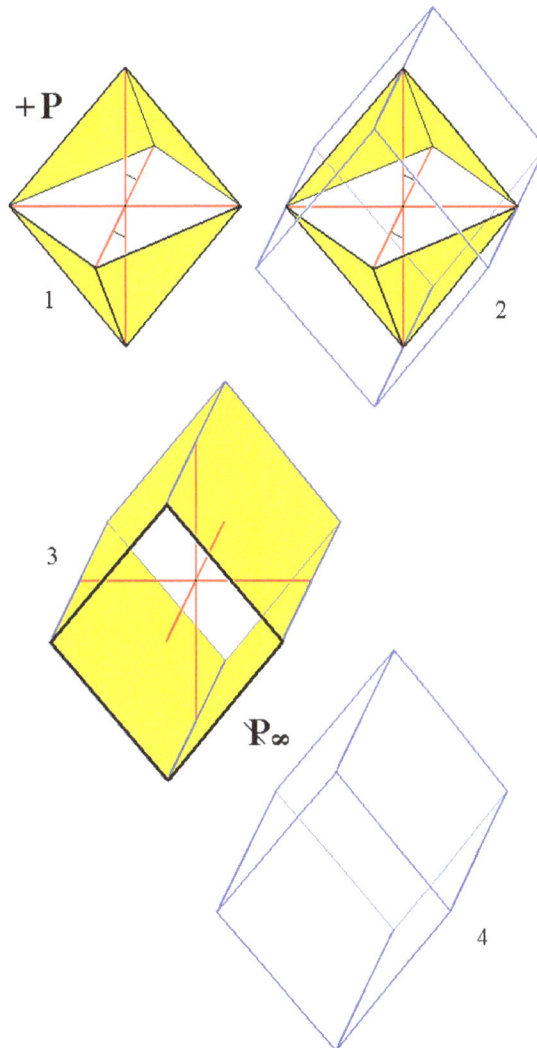

Construction (derivation) of the primary Clinodome from the primary positive Protohemipyramid.

(1). The primary positive Protohemipyramid. (2). Construction of the primary Clinodome. (3). Result of the construction. (4). Crystallographic axes removed. From any Protohemipyramid (i.e. from any member of the Vertical Series of Hemipyramids) and indeed from any Hemipyramid whatsoever (Vertical Series, Clino- or Orthodiagonal Series) a corresponding Clinodome can be derived.

From the protohemipyramid can also be derived the orthohemidome by letting the derivation coefficient n, when it refers to the ortho axis (and which is equal to 1 in all the protohemipyramids), become infinite. Then the two upper faces of the hemipyramid become one face parallel to the ortho axis (and intersecting the vertical axis), while the two lower faces of that same hemipyramid also become one face parallel to and opposite of the first face. So the result is a face pair consisting of an upper and lower face parallel to the ortho axis. If we had -- for the derivation -- taken a positive hemipyramid (+mP), then the resulting new Form consists of two parallel faces, one lower front face and one upper back face. And if we had also taken the corresponding negative hemipyramid (-mP) for such a derivation, then we would get a face pair also consisting of two parallel faces, but of which one is a lower back face and the other an upper front face. If we combine these two face pairs, then we get a horizontal prism, parallel to the ortho axis, an orthodome. So this orthodome is not a simple Form, it is a combination of two orthohemidomes a positive one and a negative one. The Weissian symbol for the orthohemidome is (a : ~b : mc) [more specifically, it is the symbol for the negative hemidome -- for the positive hemidome it is (a : ~b : -mc)], its Naumann symbol is $\pm\, m\,\mathbf{P}\,\infty$, and the Miller symbol for the positive hemidome is {hol*}, for the negative {hol}. The + sign in the Naumann symbol refers to that orthohemidome of which the faces lie in the acute angle beta, i.e. the upper back and lower front faces. With this we've found yet another basic facecompatible with the Monoclinic Crystal system, namely   a : ~b : mc. With the monoclinic orthohemidome we have found yet another Form for our list of monoclinic holohedric Forms. In the next figure we will derive the primary (m = 1) positive orthohemidome from the primary positive protohemipyramid (+P).

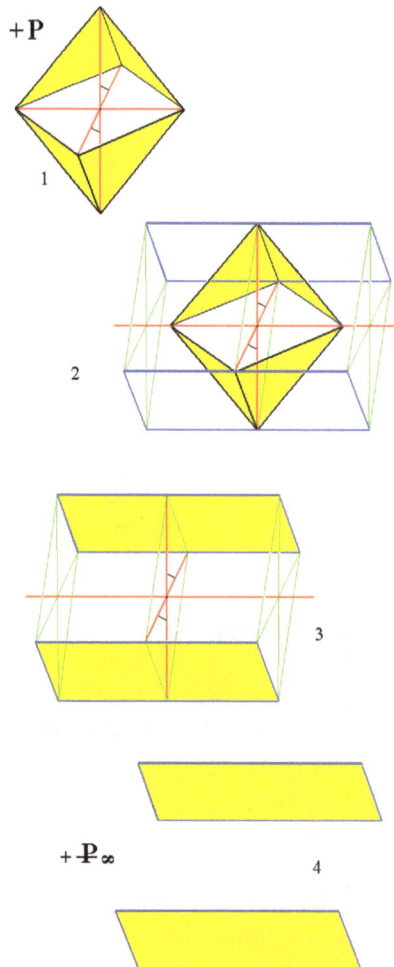

(1). The primary positive Monoclinic Protohemipyramid.

(2). Construction (derivation) of the primary positive Monoclinic Orthohemidome from the primary positive Monoclinic Protohemipyramid. The green lines are just visual aids.

(3). The resulting primary positive Monoclinic Orthohemidome, with crystallographic axes.

(4). Crystallographic axes and visual aids removed. From any positive Monoclinic Protohemipyr-amid can be constructed a corresponding positive Monoclinic Orthohemidome. Indeed from any Hemipyramid can be derived a corresponding Orthohemidome.

The negative orthohemidome can be derived from the negative protohemipyramid :

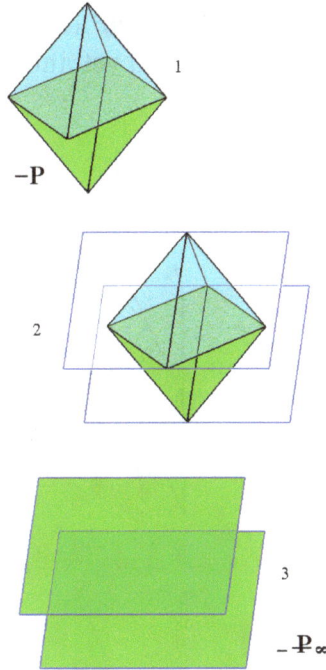

(1). The primary negative Monoclinic Protohemipyramid.

(2). Construction (derivation) of the primary negative Monoclinic Orthohemidome from the pri-mary negative Monoclinic Protohemipyramid.

(3). Final result of the construction. From any negative Monoclinic Protohemipyramid can be con-structed a corresponding negative Monoclinic Orthohemidome. And as has been said, from any Hemipyramid can be derived a corresponding Orthohemidome.

When we combine the two constructed primary orthohemidomes we will get an orthodome.

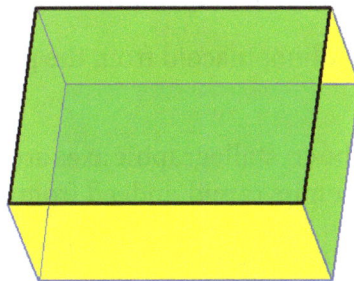

$$+ P_\infty \, . - P_\infty$$

The primary Orthodome, as a combination of a positive and negative
primary Orthohemidome. Any two corresponding positive and negative Orthohemidomes
can combine to form a corresponding Orthodome.

From the monoclinic protohemipyramid we can also derive the clinopinacoid when we let the derivation coefficient referring to the clino axis, as well as the one referring to the vertical axis become infinite. We then obtain a pair of faces parallel to the clino and vertical axes. It can close the orthodome at its left and right sides. The Weissian symbol of this Form is (~a : b : ~ c), the Naumann symbol is $\infty P\infty$, and the Miller symbol is {010}. With all this we've found yet another basic face compatible with the Monoclinic Crystal System, namely   ~a : b : ~ c. With the monoclinic clinopinacoid we have found yet another Form for our list of monoclinic holohedric Forms.

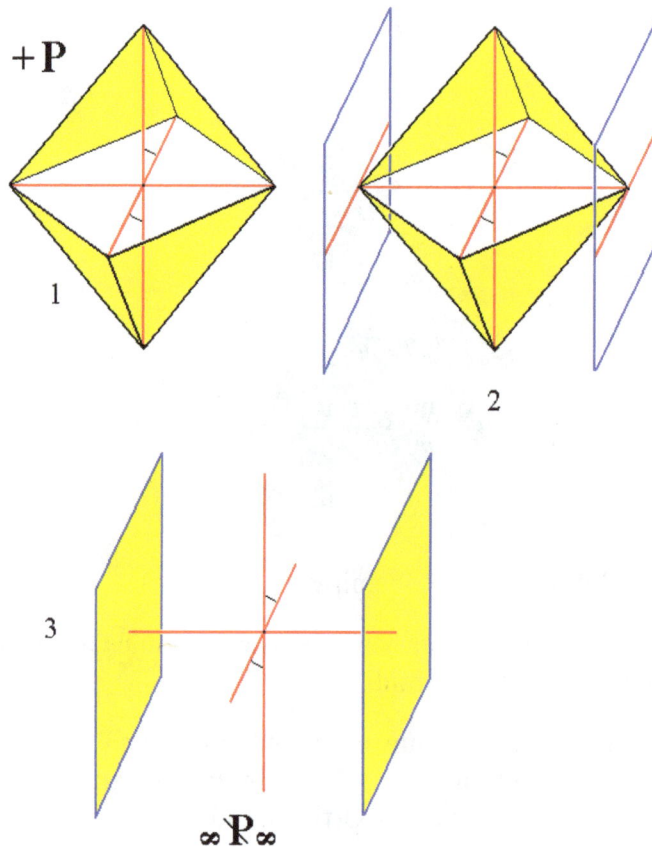

(1). The primary positive Monoclinic Protohemipyramid.

(2). Construction (derivation) of the Clinopinacoid from the primarypositive Monoclinic Protohemipyramid.

(3). The result of the construction. The crystallographic axes are shown by red lines. This Clinopinacoid can be derived from any Protohemipyramid, indeed from any Hemipyramid.

Also from any protohemipyramid can be derived the orthopinacoid, when we let the derivation coefficient referring to the ortho axis as well as the one referring to the vertical axis become infinite. It consists of a face pair parallel to the ortho and vertical axes, and could close the near and far ends of the clinodome. The Weissian symbol is (a : ~b : ~c), its Naumann symbol is $\infty P\infty$, and its Miller symbol is {100}. With all this we've found yet another basic face compatible with the

Monoclinic Crystal System, namely   a : ~b : ~c. With the monoclinic orthopinacoid we have found yet another Form for our listing of the monoclinic holohedric Forms.

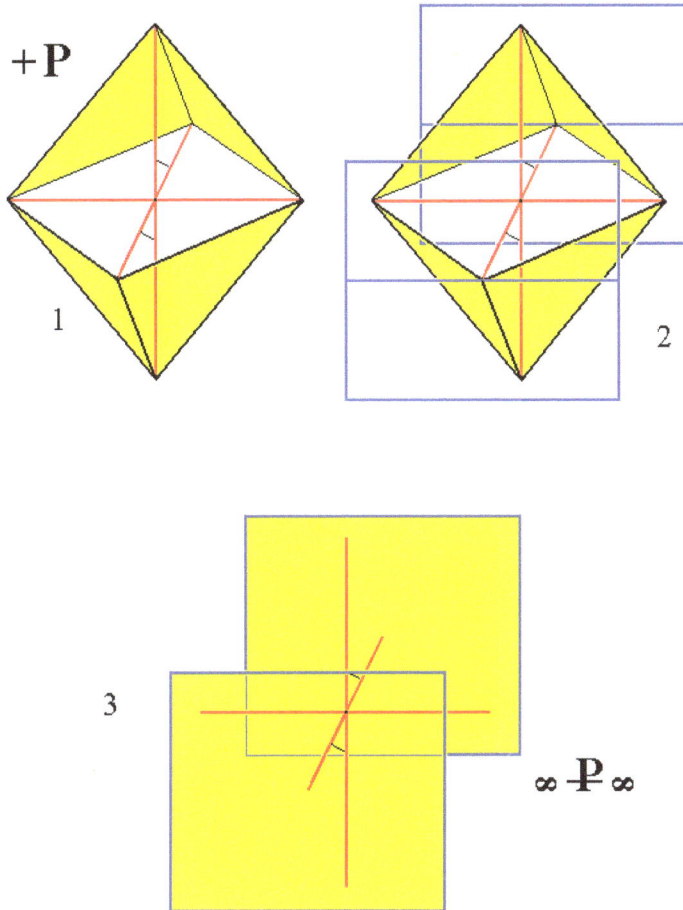

(1). The primary positive Monoclinic Protohemipyramid.

(2). Construction (derivation) of the Orthopinacoid from the primary positive Monoclinic Protohemipyramid.

(3). The result of the construction. The crystallographic axes are shown by red lines. This Orthopinacoid can be derived from any Protohemipyramid, indeed from any Hemipyramid.

Finally, we can derive the monoclinic basic pinacoid from any protohemipyramid, by letting the derivation coefficient referring to the clino axis as well as the one referring to the ortho axis become infinite. It consists of two faces parallel to the clino and ortho axes and it can close any monoclinic vertical prism at its bottom and top. The Weissian symbol of the monoclinic basic pinacoid is (~a : ~b : c), its Naumann symbol is oP, and its Miller symbol is {001}. With all this we've found yet another basic face compatible with the Monoclinic Crystal System, namely   ~a : ~b : c. With the monoclinic basic pinacoid we have found the last Form for our list of monoclinic holohedric Forms.

Remark : Instead of "basic pinacoid" we can also write "basal pinacoid".

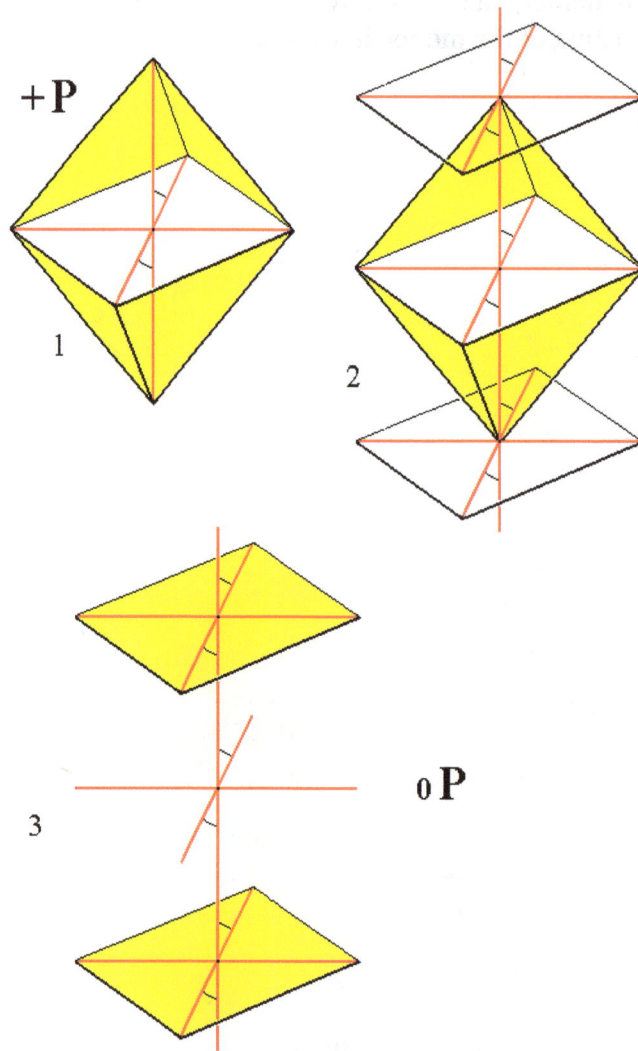

(1). The primary positive Monoclinic Protohemipyramid.

(2). Construction (derivation) of the Basic Pinacoid from the primary positive Monoclinic Proto-hemipyramid.

(3). The result of the construction. The crystallographic axes are shown by red lines in the center of the image. They are, for reasons of clarity, mirrored above and below.

This basic Pinacoid can be derived from any Protohemipyramid, and indeed from any Hemipyramid. This concludes our derivation of all the forms of the Monoclinic-prismatic Crystal Class.

Summarizing we can now list all these Forms as follows :

The Forms of the Monoclinic-prismatic Crystal Class

(= Holohedric division of the Monoclinic Crystal System)

Generalized Monoclinic Protohemipyramid.

Generalized Monoclinic Clinohemipyramid.

Generalized Monoclinic Orthohemipyramid.

Monoclinic Protoprism.

Generalized Monoclinic Clinoprism.

Generalized Monoclinic Orthoprism.

Generalized Monoclinic Clinodome.

Generalized Monoclinic Orthohemidome.

Monoclinic Clinopinacoid.

Monoclinic Orthopinacoid.

Monoclinic Basic Pinacoid.

All these Forms can (and, in this case must) enter in combinations with each other in real crystals.

Facial Approach

While deriving the above Forms from the Basic Form (or, equivalently, from its parts) we also found the corresponding basic faces compatible with the Monoclinic Crystal System. We will shortly derive those same Forms by subjecting these basic faces one by one to the symmetry elements of the present Class, the Monoclinic-prismatic Class (= holohedric division of the Monoclinic System). The derivations will be done with the aid of the stereographic projections of faces and symmetry elements. Recall that these symmetry elements were the following:

- One mirror plane.
- One 2-fold rotation axis, perpendicular to the mirror plane, and coincident with the crystallographic ortho axis.
- Center of symmetry.

From the above results we will now compose a list of all the basic faces compatible with the Monoclinic System :

   a : b : mc

   na : b : mc

   a : nb : mc

   a : b : ~c

   na : b : ~c

   a : nb : ~c

~a : b : mc

a : ~b : mc

~a : b : ~ c

a : ~b : ~c

~a : ~b : c

The stereographic projection of the symmetry elements of the present Class and of all the faces of the most general Form (a monoclinic orthohemipyramid) is given in the next Figure. Also the projections of the piercing points of the clino axis, i.e. the projections -- onto the projection plane -- of the points of intersection of the clino axis with the projection sphere, are given (Recall that the clino axis is not horizontal but tilted by the angle beta. This angle varies with the substance that is crystallized, and so does the location of the piercing points).

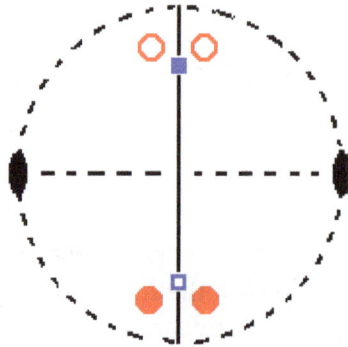

Stereogram of the symmetry elements of the Monoclinic-prismatic Crystal Class,
and of all the faces of the most general Form.

There is one mirror plane (solid line) and one 2-fold rotation axis (indicated by a pair of small solid ellipses) perpendicular to it. The projection of the piercing points of the clino axis (= a axis) is given in blue : the lower piercing point is represented by an open square, the upper piercing point by a solid square.

Stereographic projection of monoclinic crystals. With the Monoclinic System we - for the first time -- have to do with an oblique system, which here means that when we put one axis in a vertical direction then either the other two axes are perpendicular to it but their directions are non-orthogonal (not 90°) with respect to each other, or one of the axes is not orthogonal with that vertical axis, (these two possibilities) depending on which axis is chosen to be the vertical one. Here we choose the orientation of a crystal belonging to the present Class (2/m) such that the ortho axis (the axis that is perpendicular to both the other two axes) coincides with the 2-fold rotation axis, implying that the mirror plane is oriented vertically. A second axis is now oriented vertically such that the front end (positive end) of the third -- inclined -- axis is pointing obliquely downward (and the negative end obliquely upward). That second axis is now called the vertical axis or c axis, while the third axis is called the clino axis.

In the next Figure a vertical cross section through a monoclinic crystal is depicted and also the clino and vertical crystallographic axes. The face bp is parallel to the clino and ortho axes. Together with its opposite counter part it forms the monoclinic basic pinacoid. The face op is parallel to the ortho and vertical axes, and together with its opposite counterpart it forms the monoclinic ortho-pinacoid.

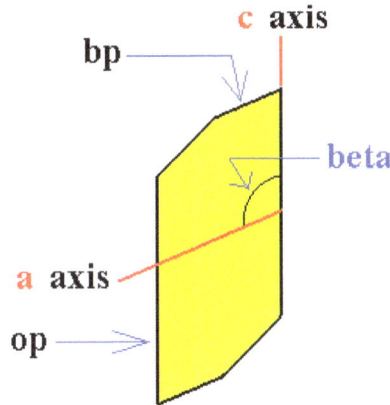

Vertical cross section through a monoclinic crystal.

The directions of the clino and vertical axes are indicated as well as the angle beta between them. The direction of the ortho axis is perpendicular to the plane of the drawing. Two faces are indicated, bp and op. The elongated direction of the crystal is chosen as the direction of the c axis (vertical axis).

We will explain some features of the stereographic projection of monoclinic crystals by considering the stereographic projection of some elements of the above Figure, namely the clino axis (a axis), the vertical axis (c axis), the face bp and the face op.

The next Figure is a vertical section through the projection sphere along the a - c plane (which means that this section contains the clino axis and the vertical axis). The construction of (1) the stereographic projection of the piercing points of the clino axis (i.e. of the points of intersection of the clino axis with the projection sphere), and (2) of the stereographic projection of the face poles of the bp and op faces is shown.

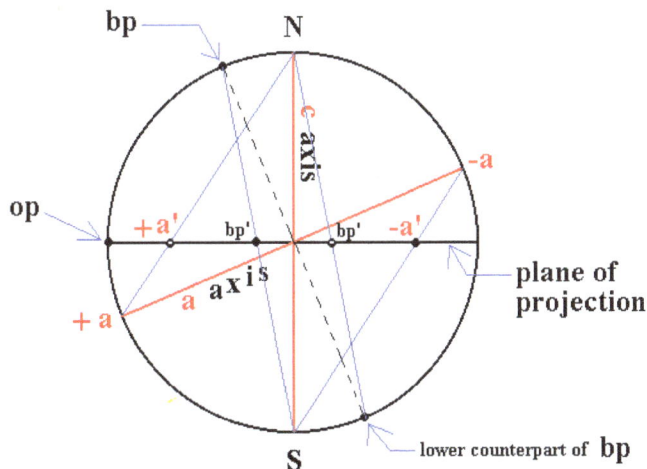

Vertical section through the projection sphere showing the constuction of the stereographic projection of some elements in Figure. Indicated are the north pole (N) and the south pole (S) of the projection sphere and the trace of the projection plane. Further we see the vertical axis (c axis) and the clino axis (a axis). The piercing points of the vertical axis (through the projection sphere) coincide with the north and south pole. The piercing points of the clino axis are indicated as +a and -a. The stereographic projections of the latter are indicated as +a' and -a'.

The projection of the face bp yields the corresponding face pole on the projection sphere (this face pole is in fact the piercing point of the line, starting from the center of the projection sphere and going in a direction perpendicular to the face bp, through the projection sphere). This face pole must now be projected onto the projection plane (the trace of which is indicated by the solid horizontal diameter in the Figure), resulting in the point bp'. The stereographic projection of the pole of the lower counter face of bp (these two faces together making up the basic pinacoid) is also indicated (as bp'). So the stereographic projection of the monoclinic basic pinacoid, which consists of two face bp, shows up as a set of two points not coincident with the center of the projection (as we were used to in the Tetragonal, Hexagonal and Orthorhombic Systems). The face pole of the face op is indicated. This face is parallel to the vertical and ortho axes. Together with its parallel counter face it forms the orthopinacoid. The stereogaphic projection of the poles onto the projection plane ends up on the perimeter of the projection sphere. One is indicated as op (in this case the face pole coincides with its stereographic projection). The other is at the opposite end of the horizontal diameter.

The next Figure depicts the stereogram of the elements considered above (and also of the clinopinacoid).

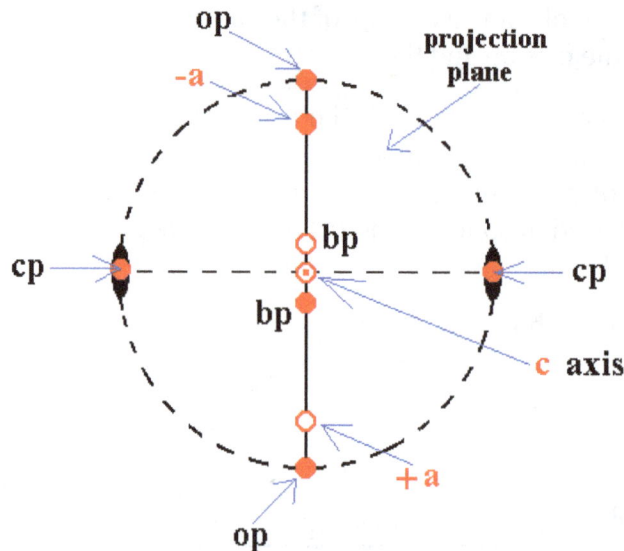

Stereographic projection of some elements discussed above, illuminating the stereographic projection of monoclinic crystals. op is the projection of the orthopinacoid, consisting of two faces parallel to the ortho and vertical axes. cp is the projection of the clinopinacoid, consisting of two faces parallel to the clino and vertical axes. bp is the basic pinacoid (basal pinacoid), consisting of two faces parallel to the clino and ortho axes. The piercing points of the vertical axis (c axis) are indicated. The piercing points of the clino axis (a axis) are indicated by +a and -a. The small solid ellipses indicate the 2-fold rotation axis. The vertical (i.e. vertical in the Figure) solid straight line indicates the mirror plane. Solid red dots indicate projections from above. Small red circles indicate projections from below. A small red circle centered with a red dot indicates two coinciding projections, one from below and one from above. The dashing of the perimeter of the projection plane (the large circle) indicates the absence of a horizontal mirror plane. Now we have enough knowledge to understand the stereographic projection of monoclinic crystals and Forms.

As has been said above we will now derive the Forms (already found above) by subjecting the basic faces one by one to the symmetry elements of the present Class :

From the face a : b : mc can be derived a monoclinic protohemipyramid when subjecting it to the symmetry elements of the present Class : The face is reflected in the mirror plane, yielding a face pair. This pair is then rotated by 180° about the 2-fold rotation axis resulting in another face pair opposite to the first one. These two face pairs make up the monoclinic protohemipyramid. Figure 22 depicts this stereographically. In that stereogram we see that the position of our initial face does not lie on the bisector of the relevant quadrant of the projection plane. This is because the clino axis is tilted. The unit distance relating to the ortho axis can be read off directly from the stereogram (because the ortho axis itself coincides with the E-W diameter of the projection plane) while the unit distance relating to the clino axis cannot so read off directly, because this axis itself does not coincide with the N-S diameter of the projection plane. What we see in the stereogram is not the clino axis itself but a projection of it onto the projection plane, and the same applies to the unit distance that is cut off from the clino axis. The projection of this unit distance is different (with respect to length) from the unit distance (associated with the clino axis) itself, resulting in the fact that the projection of the face pole (of the face a : b : mc) does not coincide with the bissector of the relevant quadrant of the projection plane.

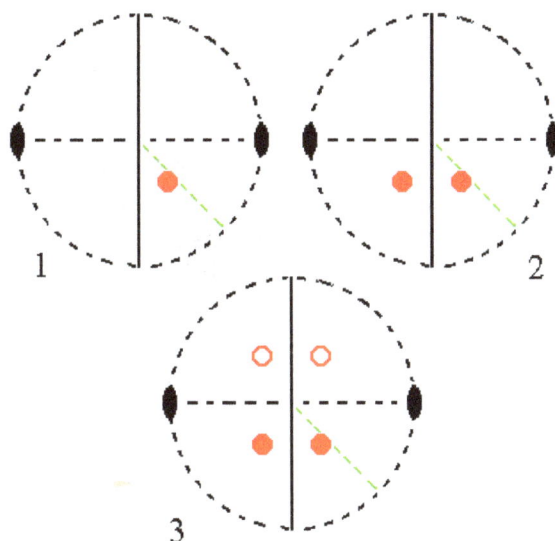

(1). Position of the face a : b : mc in the stereogram of the symmetry elements of the Monoclinic-prismatic Crystal Class.

(2). Duplication of the face in virtue of the action of the mirror plane, resulting in an upper face pair.

(3). Generation of a lower face pair by the action of the 2-fold rotation axis, resulting in four faces making up a Monoclinic Protohemipyramid (in this case a negative one).

The face na : b : mc generates a monoclinic clinohemipyramid when subjected to the symmetry elements of the present Class : The face is reflected in the mirror plane, yielding a face pair. This pair is then rotated by 180° about the 2-fold rotation axis resulting in another face pair opposite to the first one. These two face pairs make up the monoclinic clinohemipyramid. Figure depicts this stereographically.

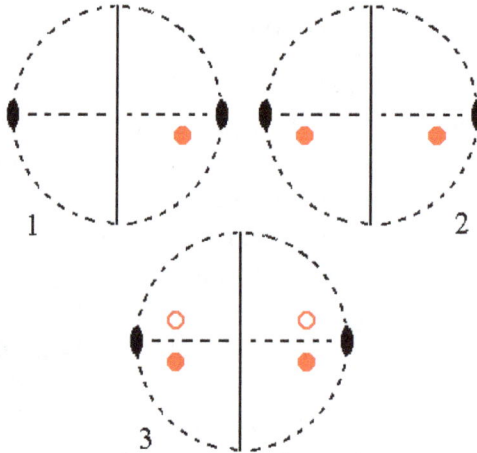

(1). Position of the face na : b : mc in the stereogram of the symmetry elements of the Monoclinic-prismatic Crystal Class.

(2). Duplication of the face in virtue of the action of the mirror plane, resulting in an upper face pair.

(3).Generation of a lower face pair by the action of the 2-fold rotation axis, resulting in four faces making up a Monoclinic Clinohemipyramid (in this case a negative one).

The face a : nb : mc generates a monoclinic orthohemipyramid when subjected to the symmetry elements of the present Class : The face is reflected in the mirror plane, yielding a face pair. This pair is then rotated by 180° about the 2-fold rotation axis resulting in another face pair opposite to the first one. These two face pairs make up the monoclinic orthohemipyramid. Figure depicts this stereographically.

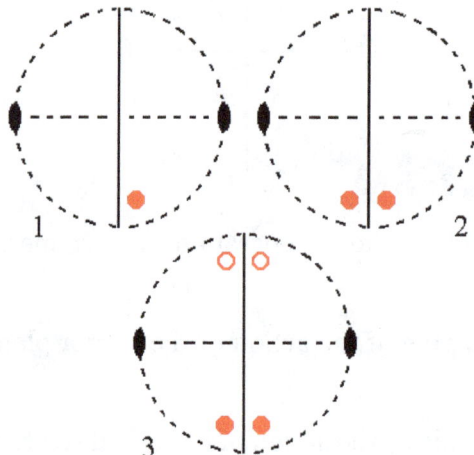

(1).Position of the face a : nb : mc in the stereogram of the symmetry elements of the Monoclinic-prismatic Crystal Class.

(2).Duplication of the face in virtue of the action of the mirror plane, resulting in an upper face pair.

(3). Generation of a lower face pair by the action of the 2-fold rotation axis, resulting in four faces making up a Monoclinic Orthohemipyramid (in this case a negative one).

The face a : b : ~c is vertical. It generates the monoclinic protoprism when subjected to the symmetry elements of the present Class :   The face is reflected in the mirror plane, yielding a vertical face pair. This pair is then rotated by 180° about the 2-fold rotation axis resulting in another vertical face pair opposite to the first one. These two face pairs make up the monoclinic protoprism. Figure depicts this stereographically.

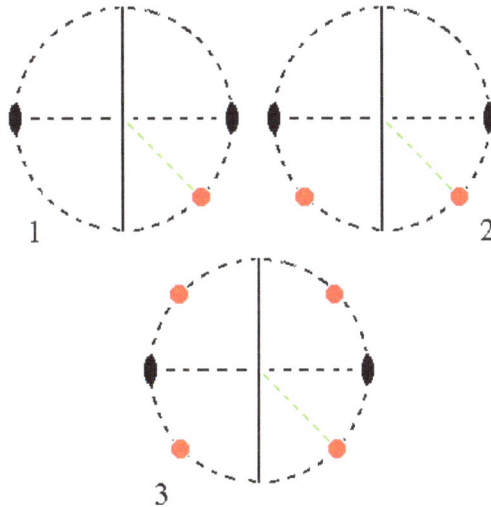

(1). Position of the face a : b : ~c in the stereogram of the symmetry elements of the Monoclinic-prismatic Crystal Class.

(2).Duplication of the face in virtue of the action of the mirror plane, resulting in a vertical face pair.

(3). Generation of an opposite vertical face pair by the action of the 2-fold rotation axis, resulting in four vertical faces making up a Monoclinic Protoprism.

In Figure we see that the projection of the face pole of the face a : b : mc does not lie on the bisector of the relevant quadrant of the projection plane, and in Figure we see that the projection of the face pole of the face a : b : ~c lies on the perimeter of the projection plane and on the bisector. With a : b : ~c we have a vertical face cutting off unit distances from the clino and ortho axes. This face can be brought into an inclined position (inclined towards the vertical axis) when we turn it about X---Y (this line serving as a hinge), such that it intersects the vertical axis at m times unit distance. We then have the face a : b : mc. Where will we find the projection of the corresponding face pole of this face on the projection plane? Figure can help us to explain that it will not lie on the bisector, even when that of the face a : b : ~c does.

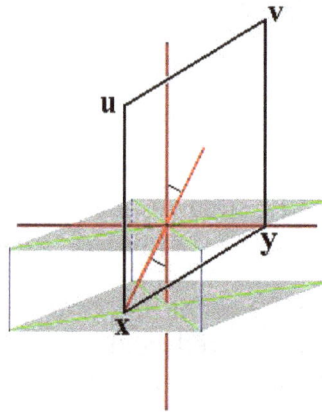

The vertical face a : b : ~c (xyuv) and its relation to the monoclinic axial system. When we turn this face
about the hinge x--y(towards the positive direction of the vertical crystallographic axis) we get
the face a : b : mc. Note that the hinge x--y is not horizontal but inclined.

When we turn the face a : b : ~c (xyuv) about the hinge x--y till it cuts m times unit distance from
the vertical axis, then we must realize that we turn it obliquely because x--y is tilted with respect to
the horizontal plane. If that face would be turned about a horizontal hinge then stereographically
we would move the projection of the corresponding face pole inwardly along the bisector.

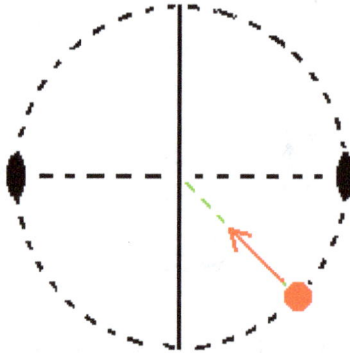

If the vertical face a : b : ~c would be turned about
a horizontal hinge its stereographic projection would move inwardly along the bisector.

But if that vertical face would (in order to generate the face a : b : mc) be turned about x--y, then it
is turned about an oblique hinge and the corresponding movement of the projection of its face pole
would be as indicated in the next Figure.

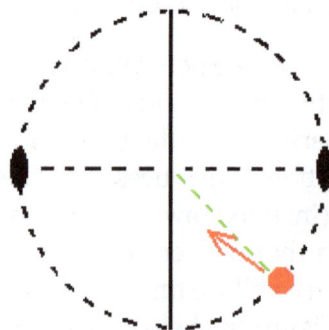

If the vertical face a : b : ~c would be turned about an oblique hinge (such as x--y)
its stereographic projection would move inwardly not along the bisector.

This is why the stereographic projection of the face a : b : mc does not lie on the bisector, even when the face a : b : ~c does so.

Remark : Of course the face a : b : ~c does, all by itself, not need to be precisely on the bisector of the relevant quadrant of the projection plane, because here we have to do with *monoclinic* crystals in which all crystallographic axes are non-equivalent. This means that the absolute cut-off distances associated with the unit face, namely one unit distance with respect to the a axis, one unit distance with respect to the b axis, and one unit distance with respect to the c axis, are not necessarily the same. In all cases where they are indeed unequal to each other this unequality will show up in the corresponding stereographic projections. But still the above described effect concerning an oblique hinge holds, causing the inclined plane to have its projected face pole shifted accordingly.

The face na : b : ~c is also vertical. It generates a monoclinic clinoprism when subjected to the symmetry elements of the present Class :   The face is reflected in the mirror plane, yielding a vertical face pair. This pair is then rotated by 180° about the 2-fold rotation axis resulting in another vertical face pair opposite to the first one. These two face pairs make up the monoclinic clinoprism. Figure depicts this stereographically.

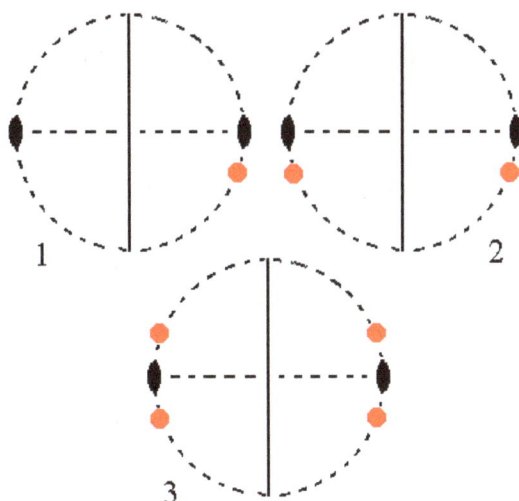

(1). Position of the face na : b : ~c in the stereogram of the symmetry elements of the Monoclinic-prismatic Crystal Class.

(2). Duplication of the face in virtue of the action of the mirror plane, resulting in a vertical face pair.

(3). Generation of an opposite vertical face pair by the action of the 2-fold rotation axis, resulting in four vertical faces making up a Monoclinic Clinoprism.

The face a : nb : ~c is also vertical. It generates a monoclinic orthoprism when subjected to the symmetry elements of the present Class: The face is reflected in the mirror plane, yielding a vertical face pair. This pair is then rotated by 180° about the 2-fold rotation axis resulting in another vertical face pair opposite to the first one. These two face pairs make up the monoclinic orthoprism. Figure depicts this stereographically.

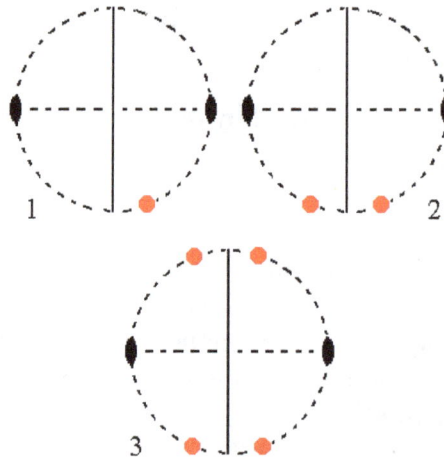

(1). Position of the face a : nb : ~c in the stereogram of the symmetry elements of the Monoclinic-prismatic Crystal Class.

(2). Duplication of the face in virtue of the action of the mirror plane, resulting in a vertical face pair.

(3). Generation of an opposite vertical face pair by the action of the 2-fold rotation axis, resulting in four vertical faces making up a Monoclinic Orthoprism.

The face ~a : b : mc is parallel to the clino axis. It generates a monoclinic clinodome when subjected to the symmetry elements of the present Class : The face is reflected in the mirror plane, yielding an upper face pair. This pair is then rotated by 180° about the 2-fold rotation axis resulting in a lower face pair obliquely beneath the first one. These two face pairs make up the monoclinic clinodome. Figure depicts this stereographically. In the stereographic projection we see that the position of our face is just below the E-W diameter of the projection plane. This is because the clino axis is tilted by the angle beta. The extent by which the projection of the face pole of our (initial) face is moved away from the E-W diagonal corresponds exactly with that angle beta.

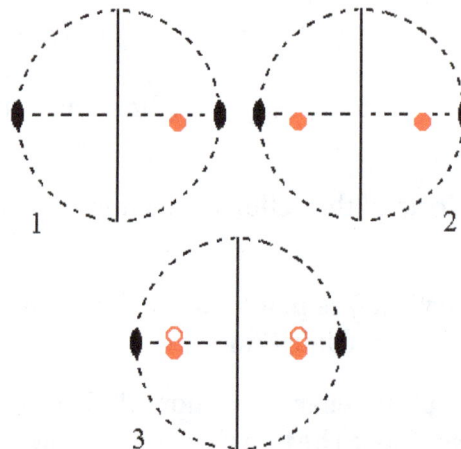

(1). Position of the face ~a : b : mc in the stereogram of the symmetry elements of the Monoclinic-prismatic Crystal Class.

(2). Duplication of the face in virtue of the action of the mirror plane, resulting in an upper face pair.

(3). Generation of a lower face pair by the action of the 2-fold rotation axis, resulting in four faces parallel to the clino axis making up a Monoclinic Clinodome. A small red circle represents a lower face.

The face a : ~b : mc is parallel to the ortho axis. It generates a monoclinic orthohemidome when subjected to the symmetry elements of the present Class: The mirror plane does not generate any new face, but the 2-fold rotation axis generates a second face opposite to the first. The resulting Form consists of two faces parallel to the ortho axis, a monoclinic orthohemidome, in the present case a negative one.

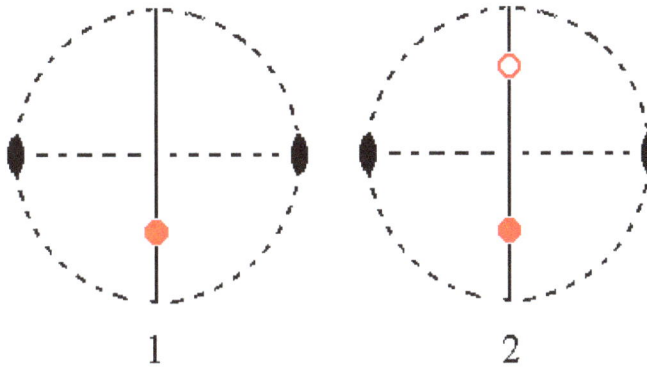

1   2

(1). Position of the face a : ~b : mc in the stereogram of the symmetry elements of the Monoclinic-prismatic Crystal Class.

(2). The mirror plane does not have any effect. The face is duplicated by the action of the 2-fold rotation axis resulting in two faces parallel to the ortho axis, but intersecting the vertical axis, a negative orthohemidome. Small red solid dots represent upper faces, small red (open) circles represent lower faces.

The positive orthohemidome can be generated from the face a : ~b : -mc, i.e. a same face as the one just used, but for which the derivation coefficient referring to the vertical axis is negative.

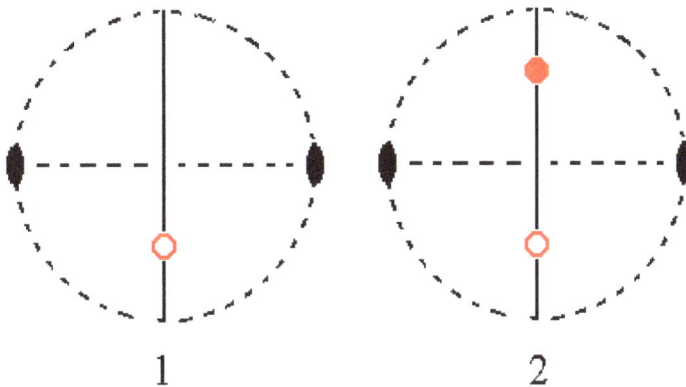

1   2

(1). Position of the face a : ~b : -mc in the stereogram of the symmetry elements of the Monoclinic-prismatic Crystal Class.

(2). The mirror plane does not have any effect. The face is duplicated by the action of the 2-fold rotation axis resulting in two faces parallel to the ortho axis, but intersecting the vertical axis, a positive orthohemidome. Combining the negative and positive Orthohemidomes yields an Orthodome, i.e. a horizontal prism, parallel to the ortho axis.

The face ~a : b : ~ c is parallel to the clino and vertical axes. It generates a monoclinic clinopinacoid when subjected to the symmetry elements of the present Class: The face is reflected in the mirror plane resulting in a vertical face pair parallel to the clino axis, the clinopinacoid.

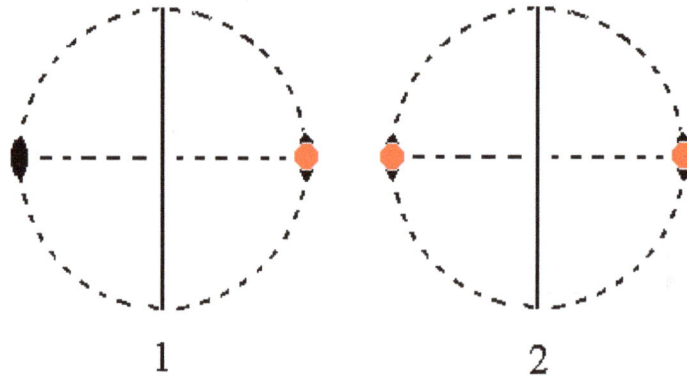

(1). Position of the face ~a : b : ~ c in the stereogram of the symmetry elements of the Monoclinic-prismatic Crystal Class.

(2). Duplication of the face by the action of the mirror plane. The 2-fold rotation axis is then implied. The resulting Form consists of two vertical faces parallel to the clino axis, the Monoclinic Clinopinacoid.

The face a : ~b : ~c is parallel to the ortho and vertical axes. It generates a monoclinic orthopinacoid when subjected to the symmetry elements of the present class: The face is rotated by 180° about the 2-fold rotation axis, resulting in a vertical face pair parallel to the ortho axis. The mirror plane is then implied. So the new Form consists of those two vertical faces parallel to the ortho axis, an orthopinacoid.

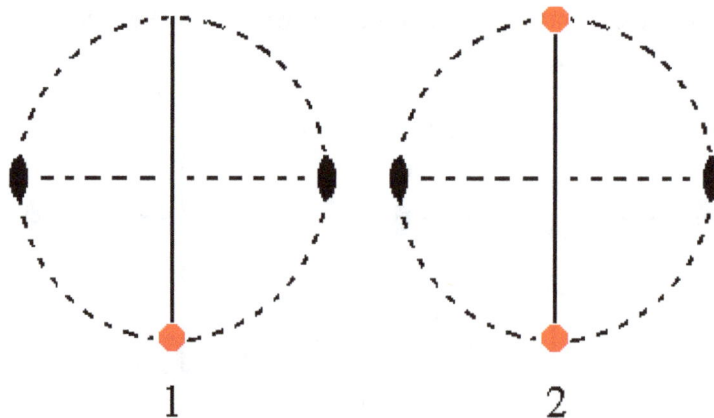

(1). Position of the face a : ~b : ~c in the stereogram of the symmetry elements of the Monoclinic-prismatic Crystal Class.

(2). Duplication of the face by the action of the 2-fold rotation axis. The mirror plane does not generate any new faces. The resulting Form consists of two vertical faces parallel to the ortho axis, the Monoclinic Orthopinacoid.

The face ~a : ~b : c is parallel to the clino and ortho axes. It is however not horizontal, but tilted, because the clino axis is not perpendicular to the vertical axis. It generates the monoclinic basic pinacoidwhen subjected to the symmetry elements of the present Class :   The face is rotated by 180° about the 2-fold rotation axis resulting in a face pair parallel to the ortho and clino axes, the basic pinacoid. The tilted orientation of the faces of the basic pinacoid is expressed in the stereographic projection by the fact that the projections of those faces do not coincide with the center of the projection plane (bounded by the (dashed) circle).

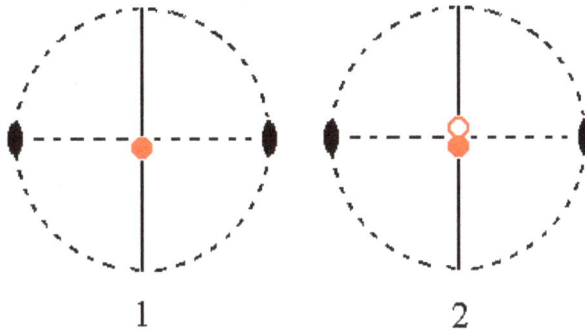

1   2

(1). Position of the face ~a : ~b : c in the stereogram of the symmetry elements of the Monoclinic-prismatic Crystal Class.

(2). Duplication of the face by the action of the 2-fold rotation axis. The mirror plane does not generate any new faces. The resulting Form consists of two faces parallel to the clino and ortho axes, the Monoclinic Basic Pinacoid.

Bravais Lattices

Two-dimensional

There is only one monoclinic Bravais lattice in two dimensions: the oblique lattice.

Three-dimensional

Two monoclinic Bravais lattices exist: the primitive monoclinic and the base-centered monoclinic lattices.

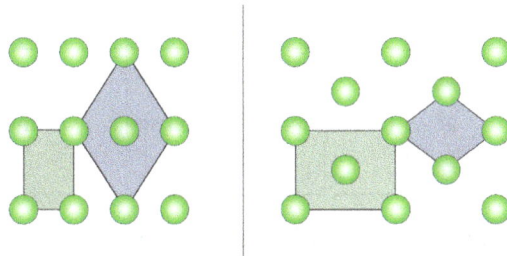

Rectangular vs rhombic unit cells for the 2D base layers of the monoclinic lattice.
The two lattices swap in centering type when the axis setting is changed

Bravais lattice	Primitive monoclinic	Base-centered monoclinic
Pearson symbol	mP	mS
Standard unit cell	$\beta \neq 90°$ $a \neq c$	$\beta \neq 90°$ $a \neq c$
Oblique rhombic prism unit cell	$\alpha \neq 90°$	$\alpha \neq 90°$

In the monoclinic system there is a rarely used second choice of crystal axes that results in a unit cell with the shape of an oblique rhombic prism; this is because the rectangular two-dimensional base layers can also be described with rhombic axes. In this axis setting, the primitive and base-centered lattices swap in centering type.

Crystal Classes

The *monoclinic crystal system* class names, examples, Schoenflies notation, Hermann–Mauguin notation, point groups, International Tables for Crystallography space group number, orbifold, type, and space groups are listed in the table below.

#	Point group					Type	Example	Space groups	
	Name	Schön.	Intl	Orb.	Cox.			Primitive	Base-centered
3–5	Sphenoi-dal	C_2	2	22	+	enantiomorphic polar	halotrichite	P2, P2$_1$	C2
6–9	Domatic	C_s (C_{1h})	m	*11	[ ]	polar	hilgardite	Pm, Pc	Cm, Cc
10–12	Prismatic	C_{2h}	2/m	2*	[2,2$^+$]	centrosymmetric	gypsum	P2/m, P2$_1$/m	C2/m
13–15								P2/c, P2$_1$/c	C2/c

Sphenoidal is also monoclinic hemimorphic; Domatic is also monoclinic hemihedral; Prismatic is also monoclinic normal.

The three monoclinic hemimorphic space groups are as follows:

- A prism with as cross-section wallpaper group p2.

- Ditto with screw axes instead of axes.

- Ditto with screw axes as well as axes, parallel, in between; in this case an additional translation vector is one half of a translation vector in the base plane plus one half of a perpendicular vector between the base planes.

The four monoclinic hemihedral space groups include

- Those with pure reflection at the base of the prism and halfway.

- Those with glide planes instead of pure reflection planes; the glide is one half of a translation vector in the base plane.

- Those with both in between each other; in this case an additional translation vector is this glide plus one half of a perpendicular vector between the base planes.

Triclinic Crystal System

In crystallography, the triclinic crystal system is one of the 7 crystal systems. A crystal system is described by three basis vectors. In the triclinic system, the crystal is described by vectors of unequal length, as in the orthorhombic system. In addition, none of the three vectors are orthogonal to another. The triclinic lattice is the least symmetric of the 14 three-dimensional Bravais lattices. It has (itself) the minimum symmetry all lattices have: points of inversion at each lattice point and at 7 more points for each lattice point: at the midpoints of the edges and the faces, and at the center points. It is the only lattice type that itself has no mirror planes. A crystal is made up of a periodic arrangement of one or more atoms (the basis) repeated at each lattice point. Consequently, the crystal looks the same when viewed from any equivalent lattice point, namely those separated by the translation of one unit cell (the motive). The volume of the unit cell can be calculated by the lattice vectors and the angles between the vectors.

$$\alpha, \beta, \gamma \neq 90°$$

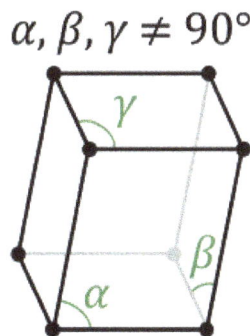

The one Basic (but complex) Form of the Triclinic Crystal System one can take the Triclinic Bipyramid, consisting of eight faces (four upper and four lower).

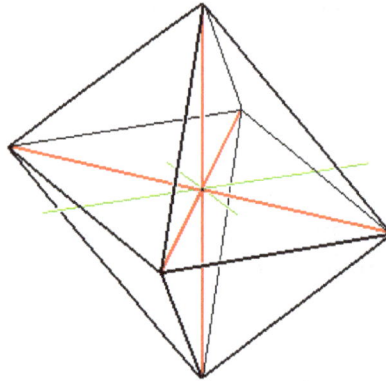

The primary Triclinic Bipyramid. It is a complex Form, that is a combination
of four simple Forms each consisting of two parallel faces.

Because such a pyramid is already a combination of four simple Forms, namely four triclinic tetar-topyramids, each consisting of two parallel faces, we can also take one of these simple Forms -- preferably the one with no minus signs, namely (a : b : c) consisting of the unit face and its parallel counter face -- as the Basic Form of the Triclinic System, from which (a set of) all other Form types can be derived (From the other three Forms making up the triclinic bipyramid, the corresponding sets of all these Form types can be derived).

The next Figure shows how the triclinic bipyramid is made up of four simple Forms, i.e. the four tetartopyramids :

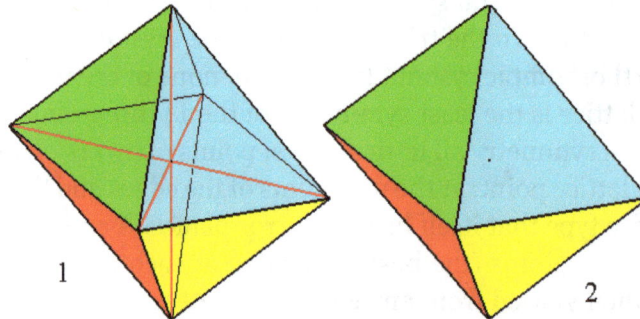

(1). The primary Triclinic Bipyramid. Each colored face represents a simple
Form (consisting of that face and its parallel counter face). (2). Crystallographic axes
and all the edges at the back side of the pyramid removed from the drawing.

All the Forms of the System's highest symmetrical Class (the Holohedric Division of our System) consist of just two parallel faces. This is so because the only symmetry element of (the crystals of) that Class is a center of symmetry : If we imagine a Form of this Class to have more than two faces, then, first of all, it should possess an even number of them, like 4, 6, 8, etc., in order to be consistent with having a center of symmetry. However, such an increase in the number of equivalent faces will introduce extra symmetry elements, like mirror planes or rotation axes. Indeed, in the derivation of triclinic holohedric Forms, following the Facial Approach, we'll see that all the resulting Forms consist of only two faces which are parallel to each other. Such Forms are (also) called pinacoids.

In the next Figure we once again depict the triclinic bipyramid indicating the faces that represent the four tetartopyramids with their corresponding Naumann symbols.

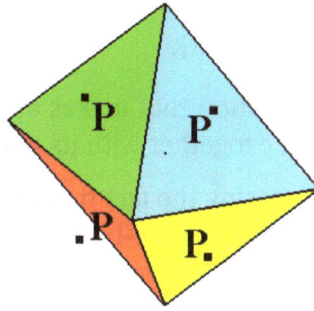

$$\text{P}^{\bullet} \cdot {}^{\bullet}\text{P} \cdot {}_{\bullet}\text{P} \cdot \text{P}_{\bullet}$$

The primary Triclinic Bipyramid. It is a complex Form, that is a combination of four simple Forms each consisting of two parallel faces.

The (simple) Forms (tetartopyramids) are denoted by their Naumann symbols. The position of the little black squares indicates the position of the face (representing the Form) on the Triclinic Bipyramid. The symbol for the combination of the four Forms making up the bipyramid is given below the drawing of that pyramid.

The Triclinic-pinacoidal Class

(= Holohedric Division of the Triclinic System) 1*

The symmetry elements of this Class are the folowing :

- Center of symmetry.

This Crystal Class shows three main types of Forms :

1. Forms, the faces of which intersect all three crystallographic axes. We call them pyramids.

2. Forms, the faces of which intersect only two crystallographic axes, the prisms and domes (each of which has its faces parallel to a third of these axes).

3. Forms, the faces of which intersect only one crystallographic axis, the pinacoids (proper) (each of which has its faces parallel to the two other axes).

Pyramids

The unit face (for the Triclinic System) a : b : c intersects all three crystallographic axes at unit distances. If we consider this face together with its parallel counter face -a : -b : -c, then we have a *face configuration* that has a symmetry consistent with that of our Class, namely just a center of symmetry. So then we have a (simple) Form. Of course it is an *open* Form and can only exist when it enters into a combination with other Forms such that a closed face configuration is the result. Our Form, consisting of the parallel face pair a : b : c and -a : -b : -c, can be

denoted by the Naumann symbol $\text{P}^{\bullet}$, expressing the fact that its front face can be considered as the upper right front face of the above mentioned triclinic bipyramid, which in turn can be considered as a *combination* of four such Forms -- tetartopyramids. The Weissian symbol for this Form is (a : b : c), and the Miller symbol is {111}.

A second Form (belonging to the combination that makes up the triclinic bipyramid) can be conceived when we consider the face a : -b : c together with its parallel counter face -a : b : -c. Its Naumann symbol can be written as $\overset{\bullet}{P}$ expressing the fact that its front face is the upper left front face of the triclinic bipyramid. Its Weissian symbol is (a : -b : c) and its Miller symbol is {11*1} (where the sign " * " stands for a minus sign placed above the preceding numeral -- as is done in the literature).

A third Form (belonging to the combination that makes up the triclinic bipyramid) can be conceived when we consider the face a : -b : -c together with its parallel counter face -a : b : c. Its Naumann symbol can be written as $_{\bullet}P$ expressing the fact that its front face is the lower left front face of the triclinic bipyramid. Its Weissian symbol is (a : -b : -c) and its Miller symbol is {11*1*}.

A fourth Form (belonging to the combination that makes up the triclinic bipyramid) can be conceived when we consider the face a : b : -c together with its parallel counter face -a : -b : c. Its Naumann symbol can be written as $P_{\bullet}$ expressing the fact that its front face is the lower right front face of the triclinic bipyramid. Its Weissian symbol is (a : b : -c) and its Miller symbol is {111*}.

The Forms we have just discussed are, however, primary Forms, meaning here that in all of them the derivation coefficient m (which refers to the vertical axis) is equal to 1. In compiling a list of holohedric Forms we will, wherever possible, strive for a listing of just all the Form types, which boils down to enumerating generalized Forms in all cases where such a generalization is possible. The above four Forms will not enter in our list because they are not generalized, but they nevertheless serve as Forms *from which* all other Forms of the Class can be derived. So let's start to generalize (which is equivalent to the finding of *types*) and begin our listing, a listing of the Form types, as well as a listing of all the single faces, the basic faces, representing those Forms.

When we let m vary, then we can discriminate three Series of (derived) tetartopyramids :

1. The Vertical Series. The faces of its Forms intersect the brachy and macro axes at the unit distances associated with them.

    These Forms are called Prototetartopyramids.

    Such a pyramid can be denoted by the Weissian symbol (a : b : mc), by the Naumann symbol $_mP^{\bullet}$, and by the Miller symbol {hhl}. Like the above case of the primary tetartopyramids, there are three more such Forms together making up the four (derived, i.e. nonprimary) prototetartopyramids :

    i.   (a : b : mc), $_mP^{\bullet}$

    ii.   (a : -b : mc), $_m\overset{\bullet}{P}$

    iii. (a : -b : -mc), $_m{}_{\bullet}P$

    iv.   (a : b : -mc), $_mP_{\bullet}$

    The first of the above four Forms will enter as the first item of our list of Form types, and the face representing it, as the first item of our list of basic faces.

2. The Brachydiagonal Series.

For the faces of its Forms the derivation coefficient m, as well as n *if* the latter refers to the brachy axis, can vary such that n is greater than 1. These Forms are called Brachytetartopyramids. In the same way as for the above mentioned pyramids there are four such pyramids :

i. $(na : b : mc)$, $_m\dot{P}\breve{n}$

ii. $(na : -b : mc)$, $_m\dot{P}\breve{n}$

iii. $(na : -b : -mc)$, $_m.P\breve{n}$

iv. $(na : b : -mc)$, $_mP.\breve{n}$

The fact that n refers to the *brachy* axis is indicated in the Naumann symbol by placing the sign ⌄ above n. The first of the above four Forms will enter as the second item of our list of Form types, and the face representing it, as the second item of our list of basic faces.

3. The Macrodiagonal Series. For the faces of its Forms the derivation coefficient m, as well as n *if* the latter refers to the macro axis, can vary such that n is greater than 1. These Forms are called Macrotetartopyramids. In the same way as for the above mentioned pyramids there are four such pyramids :

i. $(a : nb : mc)$, $_m\dot{P}\bar{n}$

ii. $(a : -nb : mc)$, $_m\dot{P}\bar{n}$

iii. $(a : -nb : -mc)$, $_m.P\bar{n}$

iv. $(a : nb : -mc)$, $_mP.\bar{n}$

The fact that n refers to the *macro* axis is indicated in the Naumann symbol by placing the sign ▬ above n. The first of the above four Forms will enter as the third item of our list of Form types, and the face representing it, as the third item of our list of basic faces.

So with these three Series we now have also found the first three of the basic faces compatible with the Triclinic Crystal System. The basic faces of the Triclinic System are such that from them can be derived the non-primary holohedric Forms by subjecting each of them to the symmetry elements of the most symmetric Class of our Crystal System (See Facial Approach, below).

To summarize, the three already found basic faces are :

a : b : mc

na : b : mc

a : nb : mc

In due course we'll find more of such basic faces.

In the next Figures we have depicted respectively the derivation of a (non primary) prototetartopyramid, a (non-primary) brachytetartopyramid, and a (non-primary) macrotetartopyramid :

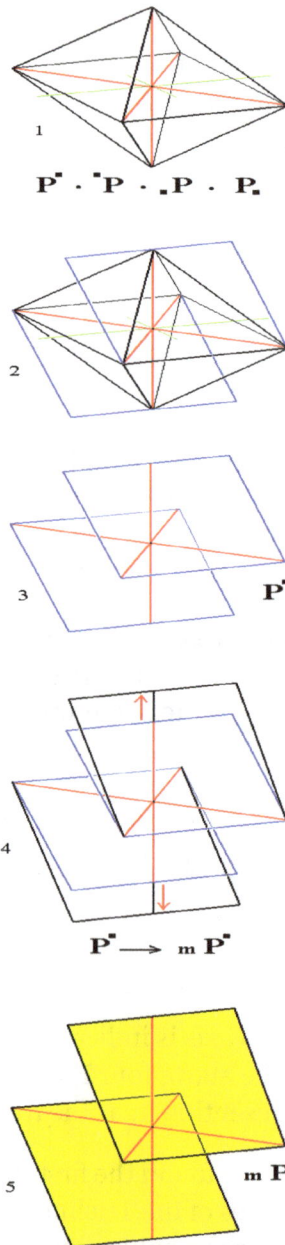

$$P^{\cdot} \cdot \overset{\cdot}{P} \cdot {}_{\cdot}P \cdot P_{\cdot}$$

$$P^{\cdot}$$

$$P^{\cdot} \longrightarrow {}_m P^{\cdot}$$

$${}_m P^{\cdot}$$

Derivation of a non-primary Prototetartopyramid.

(1). The primary Triclinic Bipyramid, a combination of four (simple) forms.

(2). Emphasis of the form (a : b : c) by extending its two faces (Extension of its faces does not change the Form crystallographically).

(3). Removal of the other Forms. The Form emphasized and isolated is the Triclinic primary Prototetartopyramid. Crystallographic axes depicted by red lines.

(4). Construction (derivation) of a Triclinic non-primary Prototetartopyramid by making m > 1

(5). Result of the construction of a Triclinic non-primary Prototetartopyramid, namely (a : b : mc).

The Form consists of two parallel faces.

Remark : In reporting on the crystallography of a new triclinic mineral or whatever new crystalline substance, or one that has not been recorded in the literature, the convention should be followed that c < a < b, in which c is the length of the vertical axis, a the length of the brachy axis, and b the length of the macro axis. In that case the Form depicted in (5) of Figure should be rotated such that the shortest axis becomes vertical. In the above case that would be the axis there signified as the brachy axis which now becomes the vertical axis. The longest axis then should become the macro axis. After having oriented the Form in that way it then is a macrotetartopyramid (instead of a prototetartopyramid).

The next Figure shows the derivation of a triclinic non-primary brachytetartopyramid.

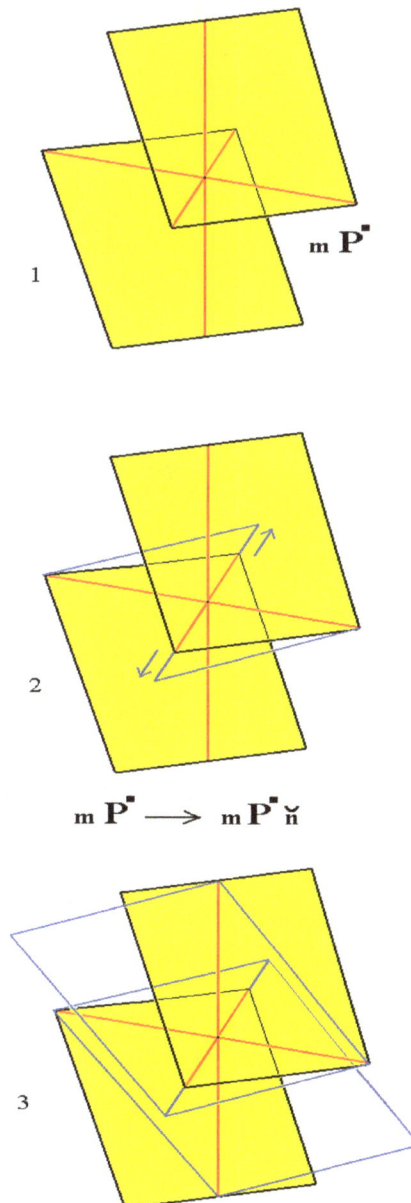

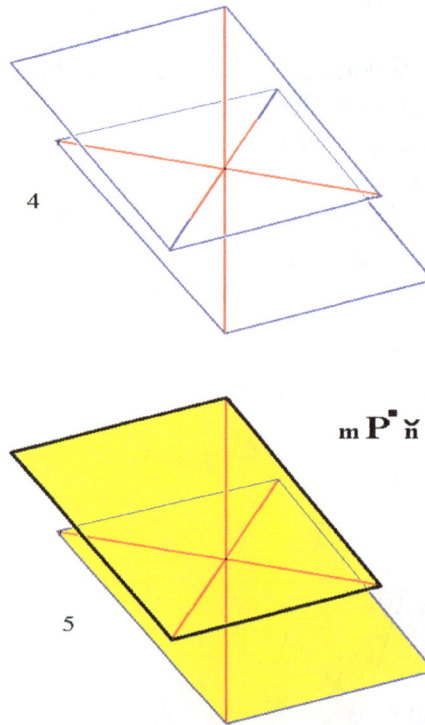

$_m P^{\bullet} \, \breve{n}$

Derivation of a Triclinic non-primary Brachy-tetartopyramid, from a Triclinic
non-primary Prototetartopyramid.

(1). A Triclinic non-primary Prototetartopyramid.

(2). Increasing the distance (measured from the origin of the axial system) of intersection of the brachy axis. The intersection distances with respect to the vertical and macro axes are held constant.

(3). Construction of the faces of the new Form.

(4). Result of the construction (derivation) of a Triclinic non-primary Brachytetartopyramid.

(5). The same as (4). Coloration applied. The three crystallographic axes are given as red lines.

The next two Figures show the derivation of a triclinic non-primary macrotetartopyramid.

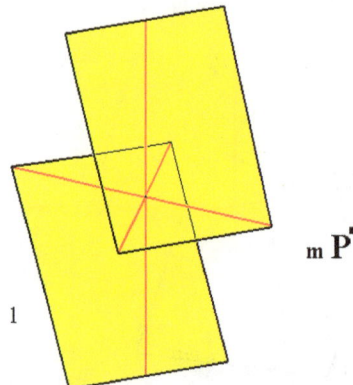

$_m P^{\bullet}$

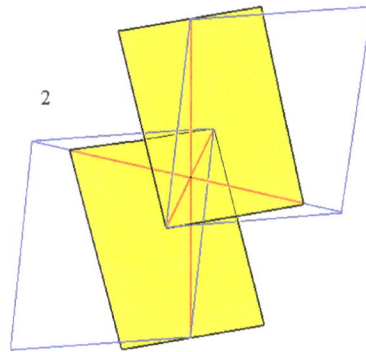

$$_m P^{\bullet} \longrightarrow \, _m P^{\bullet} \bar{n}$$

Derivation of a Triclinic non-primary Macrotetartopyramid from
a Triclinic non-primary Prototetartopyramid.

(1). A Triclinic non-primary Prototetartopyramid.

(2). Construction of the Triclinic non-primary Macrotetartopyramid, by increasing the intersection distance with respect to the macro axis. The intersection distances with respect to the brachy and vertical axes are held constant.

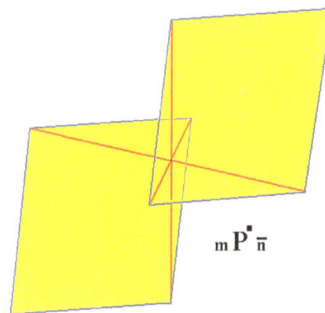

$$_m P^{\bullet} \bar{n}$$

Result of the construction, a Triclinic non-primary Macrotetartopyramid.

Prisms and Domes

These Forms can be derived from the tetartopyramids by letting one derivation coefficient become infinite. This means that their faces intersect two crystallographic axes and are parallel to a third. The prisms will be derived from the tetartopyramids by letting the derivation coefficient m become infinite resulting in their faces to be parallel to the vertical axis. Because they only consist of two (parallel) faces, they are hemiprisms. The domes will be derived from the tetartopyramids by letting the derivation coefficient n referring either to the brachy or to the macro axis become infinite, resulting in their faces to be parallel either to the brachy axis or to the macro axis. Because they consist of only two faces they are hemidomes.

Prisms

From the triclinic (tetarto-)pyramids we can derive three types of (hemi-)prism :

- The Protohemiprism, with its Weissian symbol (a : b : ~c), and its Miller symbol {110} (The Naumann symbols are given in the Figures).

- The Brachyhemiprisms, (na : b : ~c).

- The Macrohemiprisms, (a : nb : ~c).

From the triclinic pyramids of the Vertical Series we can derive the triclinic protohemiprism by letting the derivation coefficient m become infinite. Of course it does not matter whether we derive such a hemiprism from a primary (m = 1) or from a non-primary (m unequal to 1) prototetartopyramid.

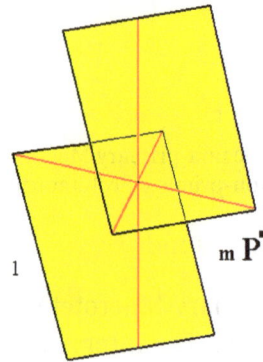

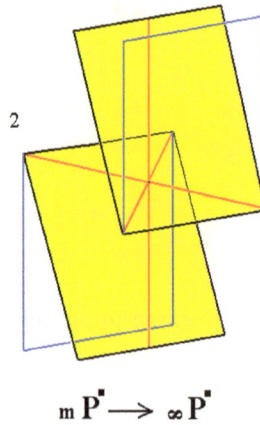

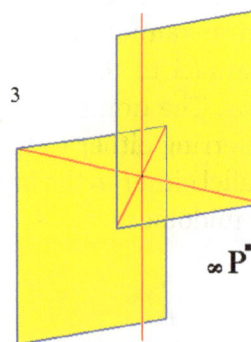

Derivation of the Triclinic Protohemiprism.

(1). The (derived) Triclinic Prototetartopyramid.

(2). Construction (derivation) of the Triclinic Protohemiprism from a Triclinic Prototetartopyramid.

(3). Result of the construction. The Form consists of two faces parallel to the vertical axis, and cutting off unit distances from the other two axes.

From the pyramids of the Brachydiagonal Series can, in the same way, be derived triclinic brachyhemiprisms, by letting the derivation coefficient m become infinite.

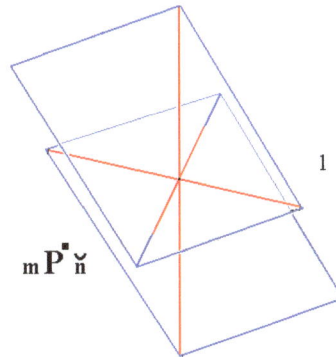

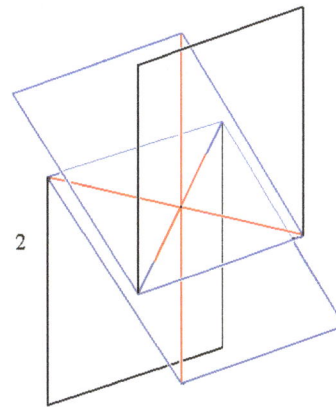

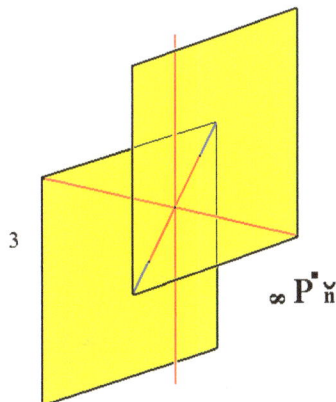

Derivation of a Triclinic Brachyhemiprism from a Triclinic Brachytetartopyramid.

(1). A Triclinic (non-primary) Brachytetartopyramid.

(2). Construction (derivation) of a Triclinic Brachyhemiprism from a Triclinic Brachytetartopyramid.

(3). Result of the construction. Again the new Form is a verical face pair.

From the pyramids of the Macrodiagonal Series can, in the same way, be derived triclinic macrohemiprisms by letting the derivation coefficient m become infinite.

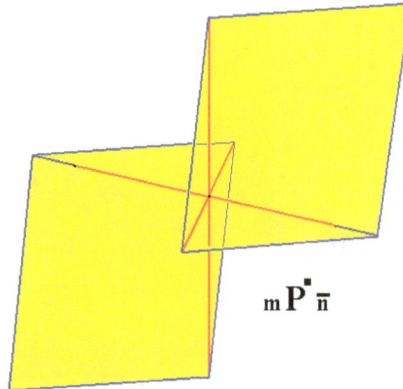

$$_m P^{\bullet} \bar{n}$$

The Triclinic (non-primary) Macrotetartopyramid, from which the
Triclinic Macrohemiprism will be derived.

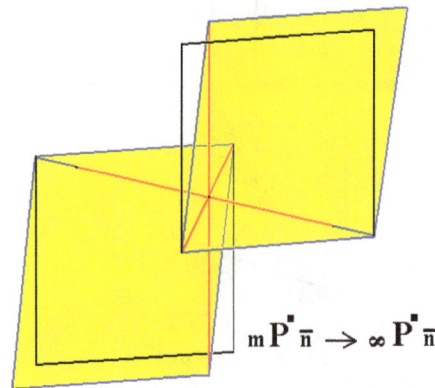

$$_m P^{\bullet} \bar{n} \rightarrow {}_\infty P^{\bullet} \bar{n}$$

Construction (derivation) of a Triclinic Macrohemiprism
from a Triclinic Macrotetartopyramid.

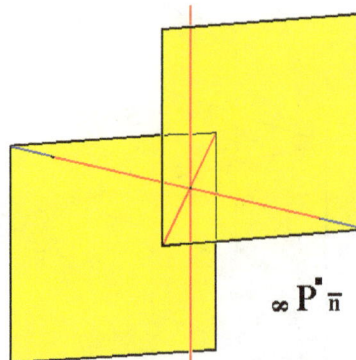

$$_\infty P^{\bullet} \bar{n}$$

Result of the construction of a Triclinic Macrohemiprism.

Above we have derived the triclinic hemiprisms from non-primary triclinic upper right tetartopyramids, i.e. from the Forms :

- $(a : b : mc)$, $_m P^{\centerdot}$

- $(na : b : mc)$, $_m P^{\centerdot} \, \check{n}$

- $(a : nb : mc)$, $_m P^{\centerdot} \, \bar{n}$

Of course those same hemiprisms can be derived from the respective *primary* versions of those tetartopyramids.

But still other triclinic hemiprisms can be derived from :

- $(a : -b : mc)$, $_m {}^{\centerdot}P$

- $(na : -b : mc)$, $_m {}^{\centerdot}P \, \check{n}$

- $(a : -nb : mc)$, $_m {}^{\centerdot}P \, \bar{n}$

I.e. from the upper left tetartopyramids.

The other tetartopyramids, the lower left and the lower right, do not yield new hemiprisms as is clear from the next Figure (in this case concerning a *primary* triclinic bipyramid, but that is immaterial for the derivation of hemiprisms).

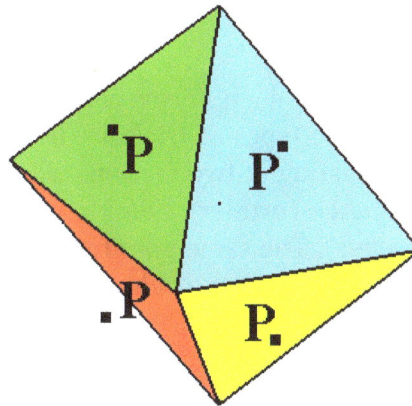

$$P^{\centerdot} \cdot {}^{\centerdot}P \cdot {}_{\centerdot}P \cdot P_{\centerdot}$$

The primary Triclinic Bipyramid. It is a complex Form, that is a combination of four simple Forms each consisting of two parallel faces.

The (simple) Forms (tetartopyramids) are denoted by their Naumann symbols. The position of the little black squares indicates the position of the face (representing the Form) on the Triclinic Bipyramid. The symbol for the combination of the four Forms making up the bipyramid is given below the drawing of that pyramid.

When we let the face, representing the Form $P^{\centerdot}$ and its parallel counter face (together making up that Form), become vertical we get the right (in contradistinction to left) protohemiprism, $\infty P^{\centerdot}$.

Exactly this same Form is obtained, however, when we let the face, representing the Form $P_\bullet$ and its parallel counter face, become vertical.

When we let the face, representing the Form $^\bullet P$ and its parallel counter face, become vertical we get the left protohemiprism, ∞P .

Exactly the same Form is obtained, however, when we let the face, representing the Form $_\bullet P$ and its parallel counter face, become vertical.

So only two protohemiprisms are possible, a right one and a left one, denoted by the Naumann symbols $\infty P^\bullet$ and ∞P .

The same applies with regard to the brachy- and macrohemiprisms, except that in their case more than one *right* and more than one *left* hemiprisms are possible because of the varying of the derivation coefficient n, referring either to the brachy axis or to the macro axis.

With the derivation of the vertical prisms we have found three more basic faces :

a : b : ~c

na : b : ~c

a : nb : ~c

Domes

Domes are in fact tilted prisms. Also here such Forms consist of only two parallel faces, which makes them hemidomes. There are two types of triclinic (hemi-)domes, namely brachyhemi-domes, (~a : b : mc), i.e. Forms consisting of two faces parallel to the brachy axis, but cutting the vertical axis either at unit distance (primary hemidomes) or at a rational multiple of it (non-primary, i.e. derived hemidomes). The same goes for the macro axis. Here we can set the face at the macro axis' unit distance and then the value for m (referring to the vertical axis) is determined as some rational multiple of its corresponding unit distance (i.e. the unit distance associated with the vertical axis). The second type of triclinic (hemi-)dome is represented by the macrohemidomes, (a : ~b : mc), i.e. Forms also consisting of two faces, but which are paral-lel to the macro axis. Like the brachyhemidomes they are not parallel to the vertical axis. When they intersect the brachy axis at unit distance then the value for m (referring to the vertical axis) is some rational multiple of its corresponding unit distance (i.e. the unit distance associated with the vertical axis).

Recall that if we have a unit face (chosen for the Crystal System concerned) that cuts off distances (measured from the origin of the axial system) a, b, c from the respective crystallographic axes, and if we also have another face that cuts off the distances p, q, r, then the derivation coefficients of that second face are p/a, q/b and r/c, where a, b and c are set as unit distances. So the derivation coefficient relating to, say, the vertical axis, namely r/c, is equal to the number of times that the unit distance (c), associated with the vertical axis, fits in the vertical axis' cutt-off distance (r) of the second face. The Miller indices are the reciprocals of the derivation coefficients reduced to whole numbers (The minus sign, where it occurs, is then placed above the relevant index. On this website we then set an asterix after and above that index). The Miller indices -- together constituting the Miller symbol -- characterizing a face are then placed between ordinary brackets. When they characterize a Form they are placed between { }.

Brachydomes

From a triclinic non-primary brachytetartopyramid can be derived a triclinic non-primary brachy-hemidome by letting the derivation coefficient n (referring to the brachy axis) become infinite.

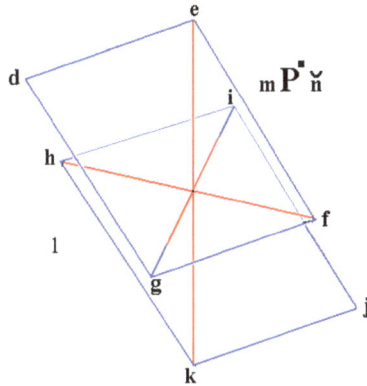

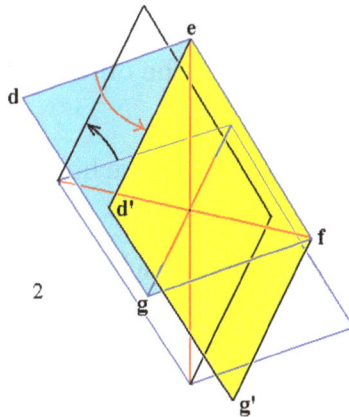

$$_{m}P^{\bullet}\check{n} \rightarrow {}_{m}P^{\bullet}\check{\infty}$$

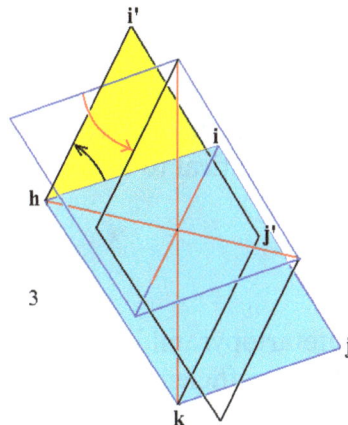

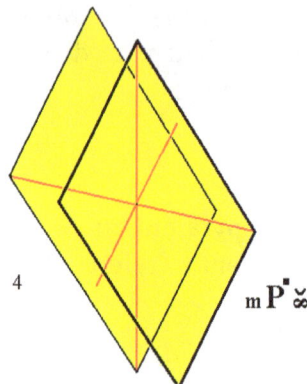

Derivation of a Triclinic (non-primary) Brachyhemidome from a Triclinic
(non-primary) Brachytetartopyramid.

(1). Two parallel faces defg and hijk making up a Triclinic (non-primary) Brachytetartopyramid.
Red lines (including blue extensions) indicate the three crystallographic axes. (2)and (3). Construction (derivation) of a Triclinic Brachyhemidome from a Triclinic Brachytetartopyramid. The
two faces become parallel to the brachy axis.

(2). Both faces of the tetartopyramid are rotated such that they become parallel to the brachy
axis. Here (in (2)) the rotation of the face defg is emphasized. It is rotated (red arrow) about
the hinge ef till it is parallel to the brachy axis, becoming the face d'efg'. It is now the front face
of the Brachyhemidome. In (3) the rotation of the other face of the tetartopyramid is emphasized.

(3). Emphasizing the rotation (black arrow) of the face hijk about the hinge hk till it is parallel to
the brachy axis. It has then become the back face, hi'j'k, of the Brachyhemidome.

(4). The result of the construction of a Triclinic Brachyhemidome. It is a Form consisting of two
faces parallel to the brachy axis, but intersecting the vertical axis and the macro axis.

Above we have constructed the brachyhemidome from a(n upper right) brachytetartopyramid,
but of course the same brachyhemidome can be derived from any (upper right) tetartopyramid.
Besides this right brachyhemidome yet another brachyhemidome can be derived, namely from
the upper left tetartopyramid. It will be the left brachyhemidome. Two such hemidomes together
make up a brachydome, i.e. an inclined prism parallel to the brachy axis. So such a prism is a combination of two (simple) Forms. The lower right tetartopyramid does not yield a new Form because
the rotation of its front face, till it is parallel to the brachy axis results in a face that is identical to
the parallel counter face of the (front face of the) left brachyhemidome. The lower left tetartopyramid also does not yield any new Form because the rotation of its front face till it is parallel to
the brachy axis results in a face that is identical to the parallel counter face of the (front face of
the) right brachyhemidome.

Macrodomes

From a triclinic non-primary macrotetartopyramid can be derived a triclinic non-primary macrohemidome, by letting the derivation coefficient n (now referring to the macro axis) become
infinite.

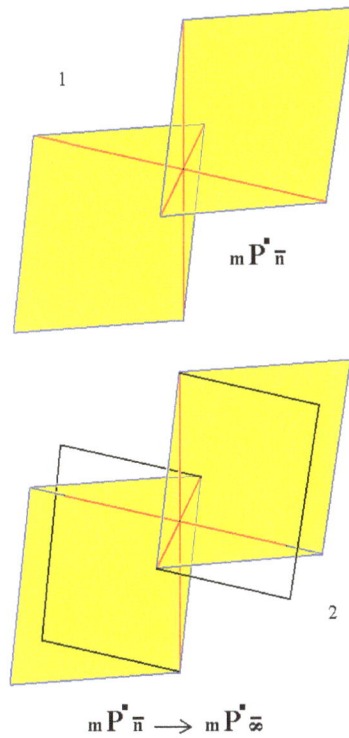

$$_m P^{\centerdot} \bar{n}$$

$$_m P^{\centerdot} \bar{n} \longrightarrow {}_m P^{\centerdot} \bar{\infty}$$

(1). A Triclinic non-primary Macrotetartopyramid.

(2). Construction (derivation) of a Triclinic non-primary Macrohemidome from the tetartopyramid of (1), by rotating the faces of the tetartopyramid about the respective hinges, one of them formed by the line connecting the intersection point on the positive vertical axis with the intersection point on the positive brachy axis, while the other hinge is formed by the line connecting the intersection point on the negative vertical axis with the intersection point on the negative brachy axis.

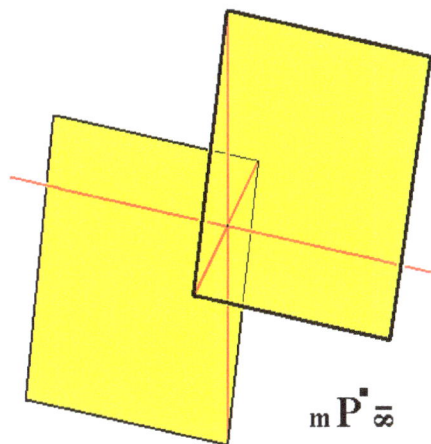

$$_m P^{\centerdot} \bar{\infty}$$

A Triclinic non-primary Macrohemidome, a Form consisting of two faces parallel to the macro axis but intersecting the brachy as well as the vertical axis.

Above we constructed a macrohemidome by rotating the faces of the tetartopyramid $P^{\centerdot}$ till they were parallel to the macro axis.

Exactly the same Form will be obtained by totating the faces of the tetartopyramid $\overset{\cdot}{\mathrm{P}}$.

The lower tetartopyramids $_\blacksquare\mathrm{P}$ and $\mathrm{P}_\blacksquare$ give identical macrohemidomes but different from those derived from the upper pyramids. So we have in fact just upper macrohemidomes (which can vary with respect to the derivation coefficient m, referring to the vertical axis) and lower macrohemidomes. Going together, a lower and an upper macrohemidome will constitute a macrodome, i.e. a combination of two (simple) Forms. Such a macrodome is an inclined prism parallel to the macro axis.

With the domes we have found yet two more basic faces :

~a : b : mc

a : ~b : mc

Pinacoids

A triclinic pinacoid is a face pair parallel to two of the three crystallographic axes, while from the other axis the faces cut off a unit distance -- in the positive or negative direction of that axis -- associated with that axis. There are three (types of) pinacoids (s. str.) depending on which axis is intersected :

- Brachypinacoid, (~a : b : ~c), resp. {010}.

- Macropinacoid, (a : ~b : ~c), resp. {100}.

- Basic Pinacoid, (~a : ~b : c), resp. {001}.

They can be derived from tetartopyramids.

Brachypinacoid

The triclinic brachypinacoid can be derived from a triclinic tetartopyramid by letting its faces become parallel to the brachy and vertical axes. This is most simply done by taking a brachyhemidome, which is itself derived from a tetartopyramid, by letting the derivation coefficient m (referring to the vertical axis) become infinite, as is done in the next Figure.

$$\mathrm{_mP^{\cdot}\underset{\infty}{\vee}}$$

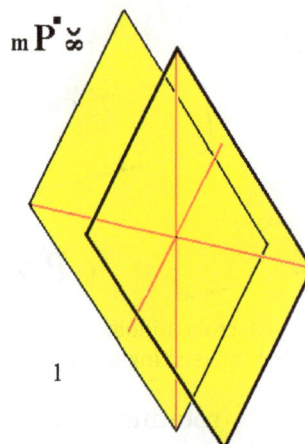

1

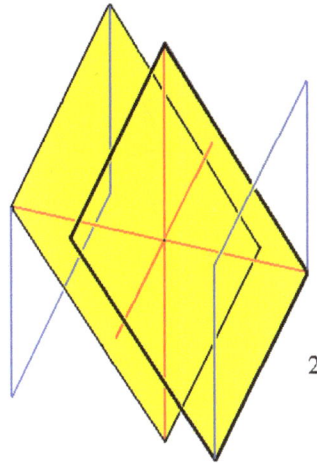

$$_m P^{\bullet} \breve{\infty} \longrightarrow {}_\infty P \breve{\infty}$$

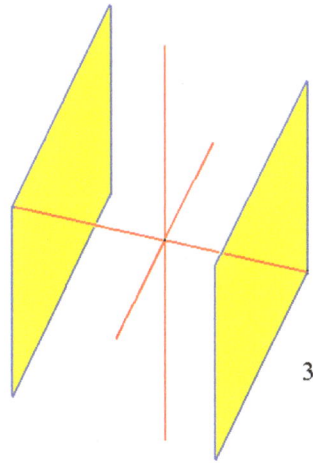

$$_\infty P \breve{\infty}$$

Derivation of the Triclinic Brachypinacoid from a Triclinic Brachyhemidome.

(1). A Triclinic non-primary Brachyhemidome.

(2). Construction (derivation) of the Triclinic Brachypinacoid from a Triclinic Brachyhemidome, by letting its faces become vertical.

(3). Result of the construction of the Triclinic Brachypinacoid.

Macropinacoid

The triclinic macropinacoid can be derived from a triclinic tetartopyramid by letting its faces become parallel to the macro and vertical axes. This is most simply done by taking a macrohemidome, which is itself derived from a tetartopyramid, by letting the derivation coefficient m (referring to the vertical axis) become infinite, as is done in the next Figure.

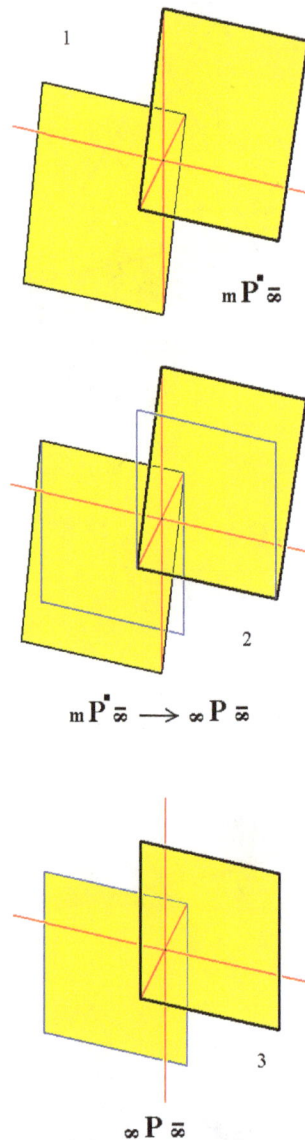

Derivation of the Triclinic Macropinacoid from a Triclinic Macrohemidome.

(1). A Triclinic non-primary Macrohemidome.

(2). Construction (derivation) of the Triclinic Macropinacoid from a Triclinic Macrohemidome, by letting its faces become vertical.

(3). Result of the construction of the Triclinic Macropinacoid. 3c. Basic Pinacoid

From any triclinic tetartopyramid, $(a : b : c)$, $(a : b : mc)$, $(na : b : mc)$, $(a : nb : mc)$, can be derived the basic pinacoid, by letting the derivation coefficient for a and for b become infinite, which implies that the two faces now are parallel to the plane in which the brachy and macro axes lie, in other words, the faces are parallel to the brachy as well as to the macro axis. Consequently the Weissian symbol for the basic pinacoid is $(\sim a : \sim b : c)$, its Naumann symbol is oP and its Miller symbol is $\{001\}$. In the next Figure we will derive the basic pinacoid from the primary prototetartopyramid, $(a : b : c)$.

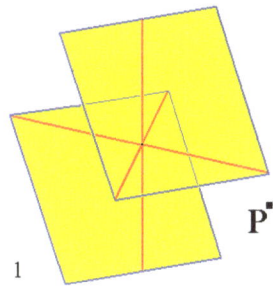

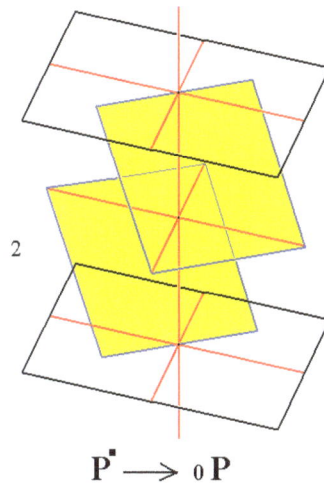

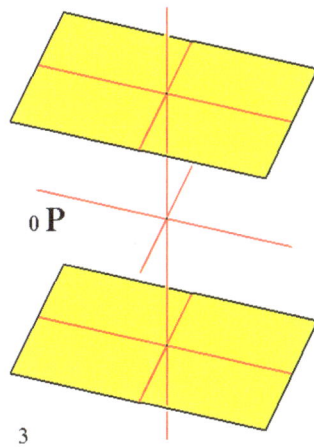

Derivation of the Triclinic Basic Pinacoid from a Triclinic Prototetartopyramid.

(1). The Triclinic primary Prototetartopyramid.

(2). Construction (derivation) of the Triclinic Basic Pinacoid from a Triclinic Prototetartopyramid, by letting its faces become parallel to the brachy and macro axes.

(3). Result of the construction of the Triclinic Basic Pinacoid.

With the pinacoids we have found the last basic faces :

~a : b : ~c

a : ~b : ~c

~a : ~b : c

This concludes our derivation of all the Forms of the Triclinic-pinacoidal Crystal Class (= Holohedric division of the Triclinic System).

All these Forms can, and in this case must, enter into combinations with each other in real crystals such that *closed* (complex) Forms appear.

So a triclinic crystal belonging to the present Class -- the System's most symmetric Class -- consists of a set of parallel face pairs, together making up a complex, closed Form.

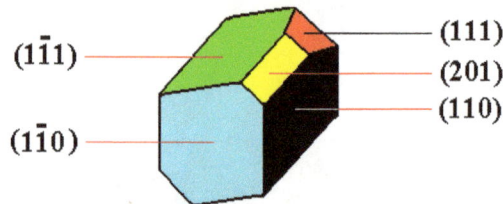

A sketch of a holohedric triclinic crystal (of the mineral Axinite).
Its faces are marked with Miller symbols.

In the above Figure we see a combination of five Forms. Each face, visible in the Figure, has a parallel counter face and as such, i.e. as parallel face pair, makes up a Form.

(110) is the front face of the right protohemiprism (a : b : ~c).

(11*0) is the front face of the left protohemiprism (which is independent of the right hemiprism) (a : -b : ~c).

(111) is the front face of the primary upper right prototetartopyramid (a : b : c).

(11*1) is the front face of the primary upper left prototetartopyramid (a : -b : c).

(201) is the front face of a non-primary upper macrohemidome (a : ~b : 2c).

With all these Forms we now also have a complete list of basic faces that we can use to again derive the Forms of the present Class as well as those of the next Class. The list reads :

a : b : mc

na : b : mc

a : nb : mc

a : b : ~c

na : b : ~c

$$a : nb : \sim c$$

$$\sim a : b : mc$$

$$a : \sim b : mc$$

$$\sim a : b : \sim c$$

$$a : \sim b : \sim c$$

$$\sim a : \sim b : c$$

Facial Approach

We will now derive those same Forms by subjecting the above listed basic faces one by one to the symmetry elements of the present Class. These symmetry elements are :

- A center of symmetry

Recall that a face configuration having a center of symmetry with respect to a certain *point*, means that the configuration is left unchanged when subjected to a reflection in that point (also called an inversion). The next Figure illustrates a face configuration, consisting of the faces A and A', having a center of symmetry (cs).

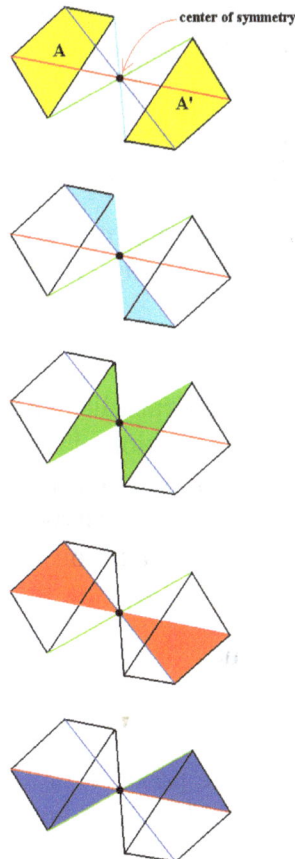

The symmetry of a face configuration, consisting of the faces A and A', consistent with a center of symmetry (cs).

The derivations of the Forms will be done with the help of stereographic projections. The next Figure is a stereographic projection of (1) the symmetry elements of the present Class, (2) the projections of all the face poles of a most general Form, a Triclinic non-primary Brachytetartopyramid, (na : b : mc), and (3) the projections of the piercing points of the brachy and macro axes (i.e. the projection onto the projection plane of the intersection points of the axes with the projection sphere).

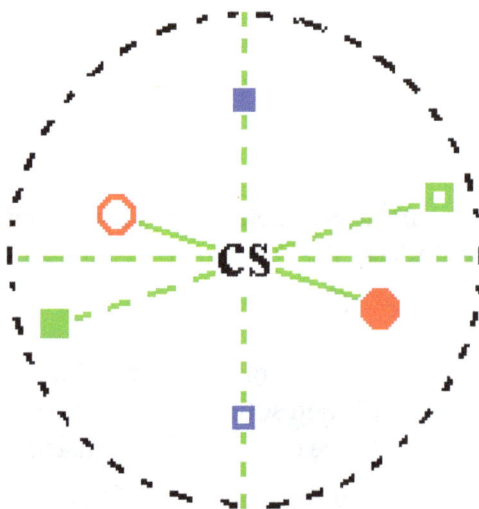

Stereographic projection of the symmetry elements (this Class having
just a center of symmetry, cs) of the Triclinic-pinacoidal Crystal Class, and of the
face poles (red dot and red circle) of a most general Form.

Blue squares are projections of the piercing points of the brachy axis (open square : projection coming from below, solid square : projection coming from above the projection plane).

Green squares are projections of the peircing points of the macro axis (The projection of the piercing points of the c axis (vertical axis) is located in the center of the projection plane (not indicated in the Figure)).

The above Figure is meant to be a very general one, i.e. us not having a specific triclinic crystal in mind. We can see that the brachy axis is not perpendicular to the vertical axis because the projections of its piercing points fall *within* the projection plane, i.e. not on its perimeter. The same is the case with the macro axis. Moreover we can see that the macro axis is not perpendicular to the brachy axis. The anglesalpha, beta and gamma between the crystallographic axes vary, their values depend on the crystallized substance under investigation.

Before we are going to derive all the Forms with the aid of the stereographic projection we'll first ponder a little more about the stereographic projection features of triclinic crystals.

Stereographic projection of triclinic crystals

In Figure we show -- approximately -- the position of all the basic faces listed above. Each of such a face is the front face of a simple Form, so we name those faces by the Forms they represent.

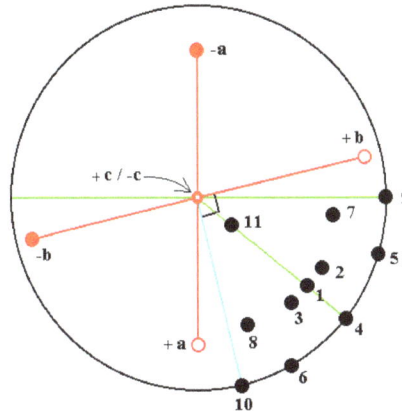

Position of the eleven Basic Faces (listed above), representing the
eleven simple Forms (Form types) :

1. Triclinic Prototetartopyramid.

2. Triclinic Brachytetartopyramid.

3. Triclinic Macrotetartopyramid.

4. Triclinic Protohemiprism.

5. Triclinic Brachyhemiprism.

6. Triclinic Macrohemiprism.

7. Triclinic Brachyhemidome.

8. Triclinic Macrohemidome.

9. Triclinic Brachypinacoid.

10. Triclinic Macropinacoid.

11. Triclinic Basic Pinacoid.

The symmetry elements are not indicated in this Figure (the only symmetry element present in crystals of this Class -- the Triclinic-pinacoidal Class -- is a center of symmetry), so one should not interpret the solid perimeter of the projection plane, drawn in this Figure, as being a horizontal mirror plane.

The brachy axis is indicated by +a -------- -a

The macro axis is indicated by +b -------- -b

7 represents a Brachyhemidome. At first one would expect it to lie on the line [+c/-c --- 9] , but because the brachy axis is inclined the face pole will come to lay somewhat beneath that line.

Something similar goes for the Macrohemidome. At first one would expect it to lie on the line [+c/-c --- 10] perpendicular to the trace of the macro axis, but because the macro axis is inclined the face pole representing the Macrohemidome will come to lie somewhat to the right of that line. The faces 11, 7 and 9 lie on the same zone parallel to the brachy axis (A zone is a set of all crystal faces, belonging to some crystal, parallel to a certain direction, say a crystallographic axis.). 11 is the basic pinacoid, it is parallel to the brachy axis but also to the macro axis. A second zone is formed by the faces 11, 8 and 10. This zone is parallel to the macro axis. The basic pinacoid not only belongs to the first zone mentioned, but also to this zone because it is parallel to the macro axis.

The face a : b : mc generates a triclinic prototetartopyramid when subjected to the symmetry elements of the present Class : The face is reflected in the center of symmetry resulting in a parallel face pair, the prototetartopyramid.

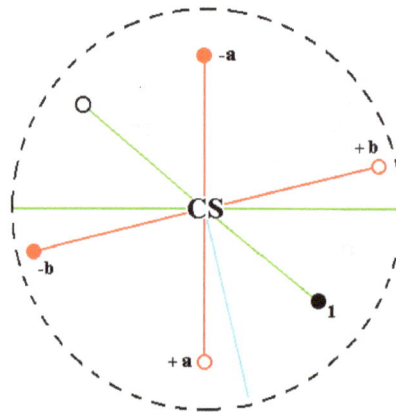

Position of the face a : b : mc (black opaque dot) in the stereogram
of the symmetry elements of the Triclinic-pinacoidal Crystal Class. The center of
symmetry generates a parallel counter face (small black open circle).

The face na : b : mc generates a triclinic brachytetartopyramid when subjected to the symmetry elements of the present Class :   The face is reflected in the center of symmetry yielding a second face parallel to the initial one. The resulting face pair is a brachytetartopyramid.

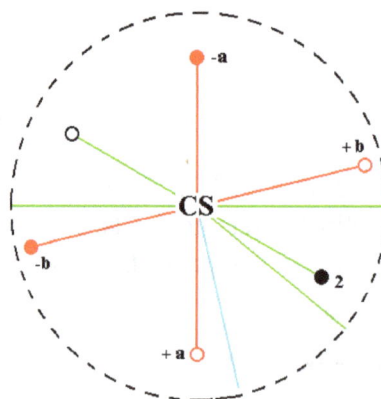

Position of the face na : b : mc (black opaque dot) in the stereogram of
the symmetry elements of the Triclinic-pinacoidal Crystal Class. The center of
symmetry generates a parallel counter face (small black open circle).

The face a : nb : mc generates a triclinic macrotetartopyramid when subjected to the symmetry elements of the present Class :   The face is reflected in the center of symmetry yielding a second face parallel to the initial one. The resulting face pair is a macrotetartopyramid.

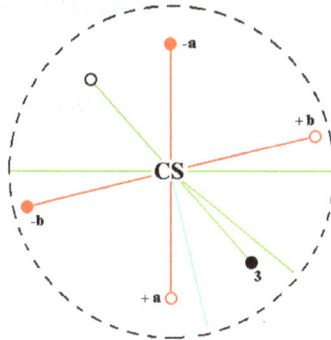

Position of the face a : nb : mc (black opaque dot) in the stereogram
of the symmetry elements of the Triclinic-pinacoidal Crystal Class. The center of
symmetry generates a parallel counter face (small black open circle).

The face a : b : ~c is vertical. It generates the triclinic protohemiprism when it is subjected to the symmetry elements of the present Class :   The face is reflected in the center of symmetry yielding a second face parallel to the initial one. The resulting face pair is a protohemiprism.

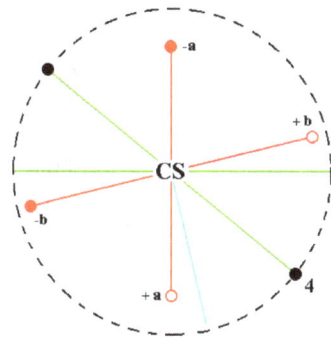

Position of the face a : b : ~c (black opaque dot in the lower right quadrant of the stereogram)
in the stereogram of the symmetry elements of the Triclinic-pinacoidal Crystal Class.

The center of symmetry generates a parallel counter face (black opaque dot in the upper left quadrant of the stereogram).

The face na : b : ~c is also vertical. It generates a triclinic brachyhemiprism when subjected to the symmetry elements of the present Class :   The face is reflected in the center of symmetry yielding a second face parallel to the initial one. The resulting face pair is a brachyhemiprism.

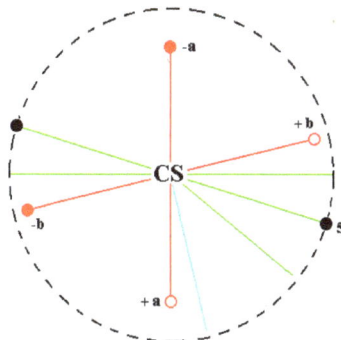

Position of the face na : b : ~c (black opaque dot in the lower right quadrant of the stereogram)
in the stereogram of the symmetry elements of the Triclinic-pinacoidal Crystal Class.

The center of symmetry generates a parallel counter face (black opaque dot in the upper left quadrant of the stereogram).

The face a : nb : ~c is also vertical. It generates a triclinic macrohemiprism when subjected to the symmetry elements of the present Class : The face is reflected in the center of symmetry yielding a second face parallel to the initial one. The resulting face pair is a macrohemiprism.

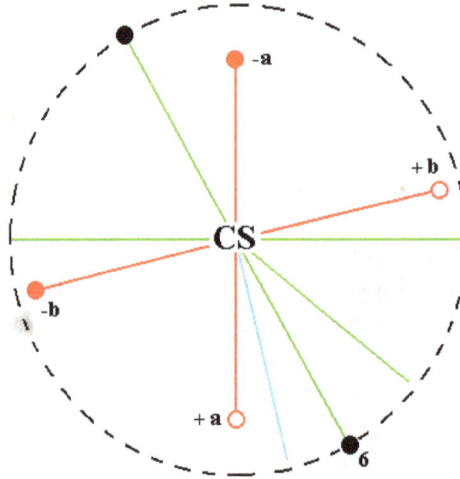

Position of the face a : nb : ~c (black opaque dot in the lower right quadrant of the stereogram) in the stereogram of the symmetry elements of the Triclinic-pinacoidal Crystal Class.

The center of symmetry generates a parallel counter face (black opaque dot in the upper left quadrant of the stereogram).

The face ~a : b : mc generates a triclinic brachyhemidome when subjected to the symmetry elements of the present Class : The face is reflected in the center of symmetry yielding a second face parallel to the initial one. The resulting face pair is a brachyhemidome.

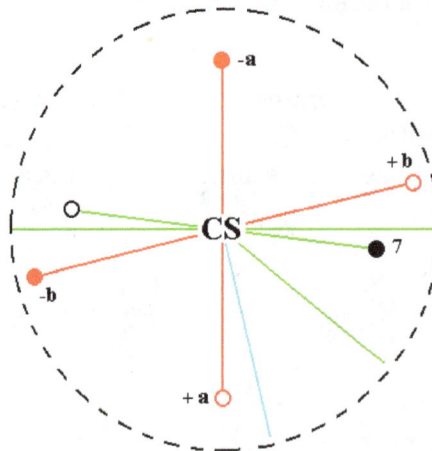

Position of the face ~a : b : mc (black opaque dot in the lower right quadrant of the stereogram) in the stereogram of the symmetry elements of the Triclinic-pinacoidal Crystal Class.

The center of symmetry generates a parallel counter face (small black open circle in the upper left quadrant of the stereogram).

The face a : ~b : mc generates a triclinic macrohemidome when subjected to the symmetry elements of the present Class : The face is reflected in the center of symmetry yielding a second face parallel to the initial one. The resulting face pair is a macrohemidome.

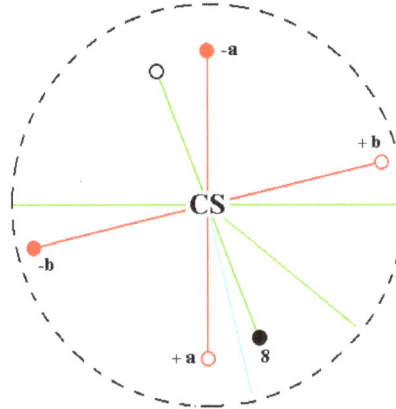

Position of the face a : ~b : mc (black opaque dot in the lower right quadrant of the stereogram) in the stereogram of the symmetry elements of the Triclinic-pinacoidal Crystal Class.

The center of symmetry generates a parallel counter face (small black open circle in the upper left quadrant of the stereogram).

The face ~a : b : ~c is vertical. It generates the triclinic brachypinacoid when subjected to the symmetry elements of the present Class : The face is reflected in the center of symmetry yielding a second face parallel to the initial one. The resulting face pair is vertical and parallel to the brachy axis, it is a brachypinacoid.

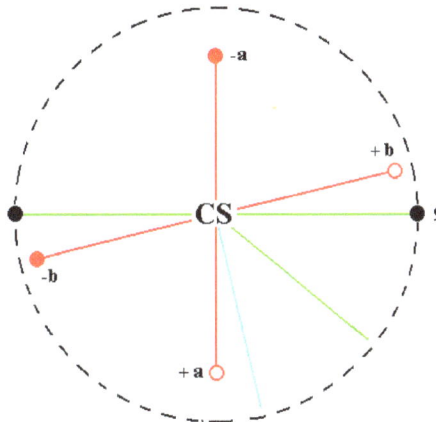

Position of the face ~a : b : ~c (black opaque dot at the right side of the stereogram) in the stereogram of the symmetry elements of the Triclinic-pinacoidal Crystal Class.

The center of symmetry generates a parallel counter face (black opaque dot at the left side of the stereogram).

The face a : ~b : ~c is also vertical. It generates the triclinic macropinacoid when subjected to the symmetry elements of the present Class : The face is reflected in the center of symmetry yielding a second face parallel to the initial one. The resulting face pair is vertical and parallel to the macro axis, it is a macropinacoid.

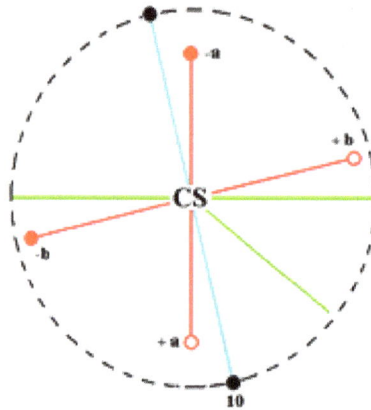

Position of the face a : ~b : ~c (black opaque dot in the lower right quadrant of the stereogram)
in the stereogram of the symmetry elements of the Triclinic-pinacoidal Crystal Class.

The center of symmetry generates a parallel counter face (black opaque dot in the upper left quadrant of the stereogram).

The face ~a : ~b : c, finally, is parallel to the brachy and macro axes. It generates a triclinic basic pinacoid when subjected to the symmetry elements of the present Class : The face is reflected in the center of symmetry yielding a second face parallel to the initial one. The resulting face pair is parallel to the brachy and macro axis, it is a basic pinacoid.

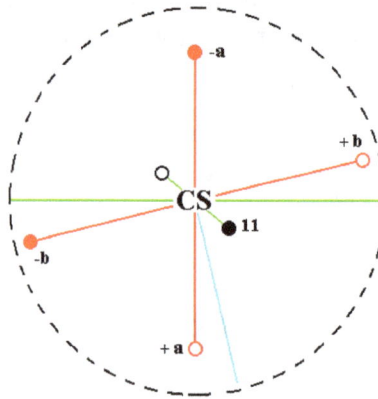

Position of the face ~a : ~b : c (black opaque dot in the lower right quadrant of the stereogram)
in the stereogram of the symmetry elements of the Triclinic-pinacoidal Crystal Class.

The center of symmetry generates a parallel counter face (small black open circle in the upper left quadrant of the stereogram).

4

Crystallographic Defects

In a crystal, the constituting atoms or molecules are positioned at fixed distances as determined by the parameters of the unit cell. The irregularities in this crystalline structure are called crystallographic defects. The aim of this chapter is to explore the different kinds of defects in crystals such as point defects, line defects, planar defects, etc.

Crystallographic deffects interrupte the regular pattern of crystal structure.

Crystal defect, imperfection in the regular geometrical arrangement of the atoms in a crystalline solid. These imperfections result from deformation of the solid, rapid cooling from high temperature, or high-energy radiation (X-rays or neutrons) striking the solid. Located at single points, along lines, or on whole surfaces in the solid, these defects influence its mechanical, electrical, and optical behaviour.

A "perfect" crystal of NaCl, for example, would consist of alternating Na+ and Cl- ions on an infinite three-dimensional simple cubic lattice, and a simple defect (a vacancy) would be a missing Na+ or Cl- ion. There are many other kinds of possible defects, ranging from simple and microscopic, such as the vacancy and other structures shown in the illustration, to complex and macroscopic, such as the inclusion of another material, or a surface.

Natural crystals always contain defects, due to the uncontrolled conditions under which they were formed. The presence of defects which affect the color can make these crystals valuable as gems, as in ruby (Cr replacing a small fraction of the Al in Al2O3). Crystals prepared in the laboratory will also always contain defects, although considerable control may be exercised over their type, concentration, and distribution.

The importance of defects depends upon the material, type of defect, and properties which are being considered. Some properties, such as density and elastic constants, are proportional to the concentration of defects, and so a small defect concentration will have a very small effect on these. Other properties, such as the conductivity of a semiconductor crystal, may be much more sensitive to the presence of small numbers of defects. Indeed, while the term defect carries with it the connotation of undesirable qualities, defects are responsible for many of the important properties of materials, and much of solid-state physics and materials science involves the study and engineering of defects so that solids will have desired properties. A defect-free silicon crystal would be of little use in modern electronics; the use of silicon in devices is dependent upon small concentrations of chemical impurities such as phosphorus and arsenic which give it desired electronic properties.

An important class of crystal defect is the chemical impurity. The simplest case is the substitutional impurity, for example, a zinc atom in place of a copper atom in metallic copper. Impurities may also be interstitial; that is, they may be located where atoms or ions normally do not exist. In metals, impurities usually lead to an increase in the electrical resistivity. Impurities in semiconductors are responsible for the important electrical properties which lead to their widespread use. The energy levels associated with impurities and other defects in nonmetals may also lead to optical absorption in interesting regions of the spectrum.

Even in a chemically pure crystal, structural defects will occur. These may be simple or extended. One type of simple defect is the vacancy, but other types exist. The atom which left a normal site to create a vacancy may end up in an interstitial position, a location not normally occupied. Or it may form a bond with a normal atom in such a way that neither atom is on the normal site, but the two are symmetrically displaced from it. This is called a split interstitial. The name Frenkel defect is given to a vacancy-interstitial pair, whereas an isolated vacancy is a Schottky defect.

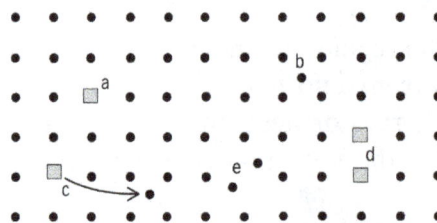

Key:
a = vacancy (Schottky defect)
b = interstitial
c = vacancy-interstitial pair (Frenkel defect)
d = divacancy
e = split interstitial
▨ = vacant site

Some simple defects in a lattice

The simplest extended structural defect is the dislocation. An edge dislocation is a line defect which may be thought of as the result of adding or subtracting a half-plane of atoms. A screw dislocation is a line defect which can be thought of as the result of cutting partway through the crystal and displacing it parallel to the edge of the cut. Dislocations are of great importance in determining the mechanical properties of crystals. A dislocation-free crystal is resistant to shear, because atoms must be displaced over high-potential-energy barriers from one equilibrium position to another. It takes relatively little energy to move a dislocation (and thereby shear the crystal), because the atoms at the dislocation are barely in stable equilibrium.

Any nonuniformity in a crystal lattice. There are four categories of crystal defects: (1) point defects, (2) line defects, (3) area defects, and (4) volume defects. Point defects include any foreign atom at a regular lattice site (substitutional site) or between lattice sites (interstitial site), anti-site defects in compound semiconductors, e.g., Ga in As or As in Ga, missing lattice atoms, and host atoms located between lattice sites and adjacent to a vacant site (Frenkel defects). Line defects, also called edge dislocations, include extra planes of atoms in a lattice. Area defects include twins or twinning (a change in crystal orientation across a lattice) and grain boundaries (a transition between crystals having no particular positional orientation to one another. Volume defects include precipitates of impurity or dopant atoms caused by volume mismatch between a host lattice and precipitates.

Point Defects

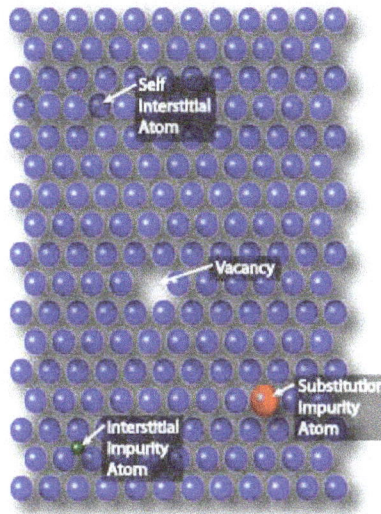

Point defects are where an atom is missing or is in an irregular place in the lattice structure. Point defects include self interstitial atoms, interstitial impurity atoms, substitutional atoms and vacancies. A self interstitial atom is an extra atom that has crowded its way into an interstitial void in the crystal structure. Self interstitial atoms occur only in low concentrations in metals because they distort and highly stress the tightly packed lattice structure.

A substitutional impurity atom is an atom of a different type than the bulk atoms, which has replaced one of the bulk atoms in the lattice. Substitutional impurity atoms are usually close in size (within approximately 15%) to the bulk atom. An example of substitutional impurity atoms is the zinc atoms in brass. In brass, zinc atoms with a radius of 0.133 nm have replaced some of the copper atoms, which have a radius of 0.128 nm.

Interstitial impurity atoms are much smaller than the atoms in the bulk matrix. Interstitial impurity atoms fit into the open space between the bulk atoms of the lattice structure. An example of interstitial impurity atoms is the carbon atoms that are added to iron to make steel. Carbon atoms, with a radius of 0.071 nm, fit nicely in the open spaces between the larger (0.124 nm) iron atoms.

Vacancies are empty spaces where an atom should be, but is missing. They are common, especially at high temperatures when atoms are frequently and randomly change their positions leaving behind empty lattice sites. In most cases diffusion (mass transport by atomic motion) can only occur because of vacancies.

Thermodynamic Aspects of Point Defects

Point defects are intentionally added to semiconductors to control the type and concentration of charge carriers. Consider, for example, boron (valence 3) as a substitutional solute in elemental silicon. The saturated covalent bonds in silicon are shown schematically in Figure, and depend on the availability of four valence electrons per silicon atom. Since the bonds are saturated, silicon has very low conductivity in its pure state; pure silicon can only conduct electricity when electrons are excited into high energy electron states. If boron is added, as in Figure, a valence electron is missing from the immediate environment of the boron atom, causing a hole in the bonding pattern. Electrons can then move from bond to bond by exchanging with the hole. The exchange requires some energy to separate the hole from the boron ion core, but this energy is small compared to that required to excite an electron from a Si-Si bond into a high-energy state. The room-temperature conductivity of Si increases significantly when a small amount of B is added. Electron-deficient solutes like boron that cause holes in the configuration of bonding electrons are called acceptors.

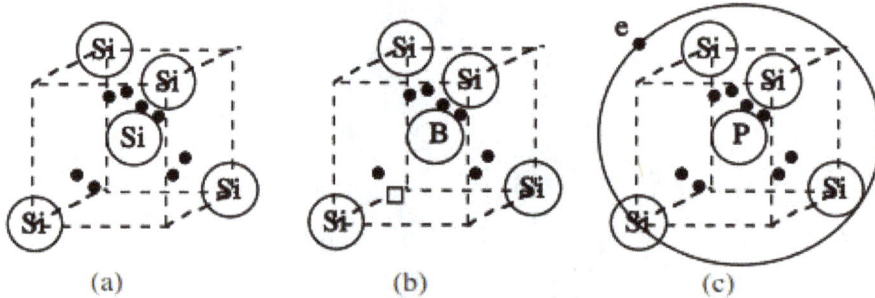

(a) Tetrahedral bonding configuration in Si. (b) Bonding around a B solute, showing a hole (□). (c) Bonding around a P solute, showing an electron in a loose orbital.

The conductivity also rises when a solute with an excess of electrons is added to a semiconductor with saturated bonds. For example, let phosphorous (valence 5) be added to Si, as in Figure. The 5 valence electrons of P are sufficient to fill the local covalent bonds with one electron left over. This electron can only go into an excited state, and orbits about the P ion core somewhat as shown in the figure. It requires a relatively small energy increment to free this electron from the P core, in which case it can transport current by moving through the lattice. The conductivity of Si rises dramatically if a small amount of P is added. Electron-excess solutes such as P in Si are called donors. Semiconductors whose electrical properties are controlled by electrically active solutes are called extrinsic semiconductors. Almost all of the semiconductors that are used in engineering devices are extrinsic.

The simplest of the point defects is a vacancy from which an atom is missing as seen in figure 7.3. All crystalline solids contain vacancies and, in fact, it is not possible to create such a material that is free of these defects. The necessity of the existence of vacancies is explained using principles of thermodynamics; in essence, the presence of vacancies increases the entropy of the crystal. The equilibrium number of vacancies for a given quantity of material depends on and increases with temperature according to

$$N_v = N_0 \exp\left(-E_v/k_B T\right)$$

In this expression, N is the total number of atomic sites, Ev is the energy required for the formation of a vacancy, T is the absolute temperature in kelvins, and k_B is the Boltzmann's constant. The value of kB is 1.38×10^{-23} J/atom-K. Thus, the number of vacancies increases exponentially with temperature; that is, as T in the above equation. For most metals, the fraction of vacancies Nv/N just below the melting temperature is on the order of 10^{-4}; that is, one lattice site out of 10,000 will be empty. A selfinterstitial is an atom from the crystal that is crowded into an interstitial site, a small void space that under ordinary circumstances is not occupied. This kind of defect is also represented in Figure. In metals, a self-interstitial introduces relatively large distortions in the surrounding lattice because the atom is substantially larger than the interstitial position in which it is situated. Consequently, the formation of this defect is not highly probable, and it exists in very small concentrations, which are significantly lower than for vacancies.

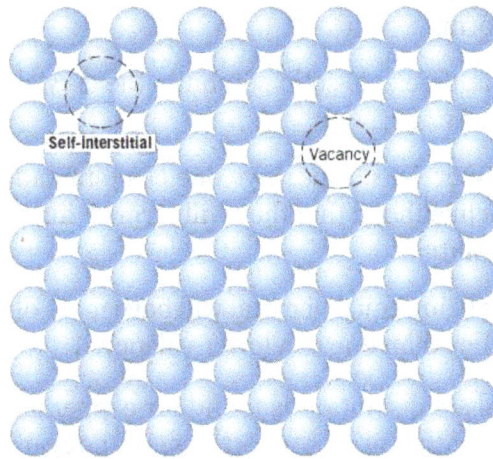

Self interstitial and vacancy in the lattice

A pure metal consisting of only one type of atom just is highly impractical; impurity or foreign atoms will always be present, and some will exist as crystalline point defects. In fact, even with relatively sophisticated techniques, it is difficult to refine metals to a purity in excess of 99.9999%. At this level, on the order of 10^{22} to 10^{23} impurity atoms will be present in one cubic meter of material. Most familiar metals are not highly pure; rather, they are alloys, in which impurity atoms have been added intentionally to impart specific characteristics to the material. Ordinarily, alloying is used in metals to improve mechanical strength and corrosion resistance. For example, sterling silver is a 92.5% silver7.5% copper alloy. In normal ambient environments, pure silver is highly corrosion resistant, but also very soft. Alloying with copper significantly enhances the mechanical strength without depreciating the corrosion resistance appreciably. The addition of impurity atoms to a metal will result in the formation of a solid solution and/or a new second phase, depending on the kinds of impurity, their concentrations, and the temperature of the alloy. Several terms relating to impurities and solid solutions deserve mention. With regard to alloys, solute and solvent are terms that are commonly employed. "Solvent" represents the element or compound that is present in the greatest amount; on occasion, solvent atoms are also called host atoms. "Solute" is used to denote an element or compound present in a minor concentration.

A solid solution forms when, as the solute atoms are added to the host material, A solid solution forms when, as the solute atoms are added to the host material, the crystal structure is maintained, and no new structures are formed. Perhaps it is useful to draw an analogy with a liquid solution. If two liquids, soluble in each other (such as water and alcohol) are combined, a liquid solution is produced as the molecules intermix, and its composition is homogeneous throughout. A solid solution is also compositionally homogeneous; the impurity atoms are randomly and uniformly dispersed within the solid.

Impurity point defects are found in solid solutions, of which there are two types: substitutional and interstitial. For the substitutional type, solute or impurity atoms replace or substitute for the host atoms. There are several features of the solute and solvent atoms that determine the degree to which the former dissolves in the latter, as follows:

1. Atomic size factor. Appreciable quantities of a solute may be accommodated in this type of solid solution only when the difference in atomic radii between the two atom types is less than about 10-12 %. Otherwise the solute atoms will create substantial lattice distortions and a new phase will form.

2. Crystal structure. For appreciable solid solubility the crystal structures for metals of both atom types must be the same.

3. Electronegativity. The more electropositive one element and the more electronegative the other, the greater is the likelihood that they will form an intermetallic compound instead of a substitutional solid solution.

4. Valences. Other factors being equal, a metal will have more of a tendency to dissolve another metal of higher valency than one of a lower valency.

An example of a substitutional solid solution is found for copper and nickel. These two elements are completely soluble in one another at all proportions. With regard to the aforementioned rules that govern degree of solubility, the atomic radii for copper and nickel are 0.128 and 0.125 nm, respectively, both have the FCC crystal structure, and their electronegativities are 1.9 and 1.8; finally, the most common valences are for copper and for nickel.

For interstitial solid solutions, impurity atoms fill the voids or interstices among the host atoms. For metallic materials that have relatively high atomic packing factors, these interstitial positions are relatively small. Consequently, the atomic diameter of an interstitial impurity must be substantially smaller than that of the host atoms. Normally, the maximum allowable concentration of interstitial impurity atoms is low (less than 10%). Even very small impurity atoms are ordinarily larger than the interstitial sites, and as a consequence they introduce some lattice strains on the adjacent host atoms. Carbon forms an interstitial solid solution when added to iron; the maximum concentration of carbon is about 2%. The atomic radius of the carbon atom is much less than that for iron: 0.071 nm versus 0.124 nm.

Point Defects are Divided into Three Types:

a. Stoichiometric Defects.

b. Impurities Defects.

c. Non-stoichiometric Defects.

Stoichiometric Defects

The compounds in which the number of positive and negative ions are exactly in the ratios indicated by their chemical formulae are called stoichiometric compounds. The defects do not disturb the stoichiometry (the ratio of numbers of positive and negative ions) are called stoichiometric defects. These are of following types,

(a) Interstitial defect:

(b) Schottky defect:

(c) Frenkel defect:

Interstitial Defect

An interstitial defect is formed when a foreign (solute) atom is positioned in the crystal structure at a point that is normally unoccupied. The defect is formed when a solute atom such as an alloying or impurity element sits within a gap between the crystal lattice points of the base metal (solvent). An interstitial atom is usually smaller than the solvent atoms located at the lattice points, but is larger than the interstitial site it occupies. Consequently, the surrounding crystal structure is distorted. An interstitial defect is often formed in metal alloys when the size of the solute atom is less than about 85% of the size of the host metal atom. In many cases, the solute atoms are less than half as small as the base metal. Carbon in iron (steel) is one example of an element that is interstitial.

Self-interstitials

Self-interstitial defects are interstitial defects which contain only atoms which are the same as those already present in the lattice.

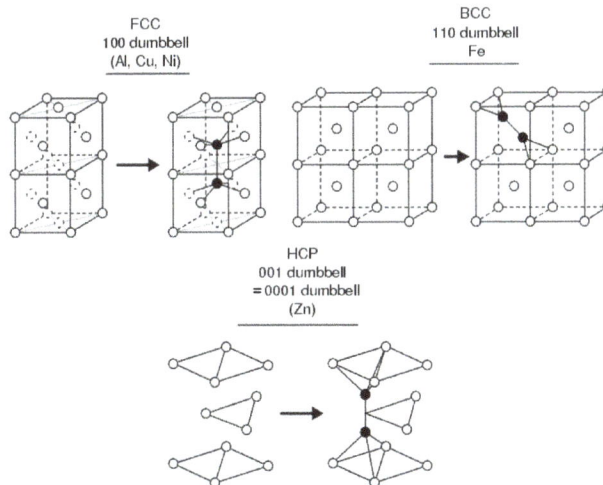

Structure of self-interstitial in some common metals. The left-hand side of each crystal type shows the perfect crystal and the right-hand side the one with a defect.

The structure of interstitial defects has been experimentally determined in some metals and semi-conductors.

Contrary to what one might intuitively expect, most self-interstitials in metals with a known structure have a 'split' structure, in which two atoms share the same lattice site. Typically the center of mass of the two atoms is at the lattice site, and they are displaced symmetrically from it along one of the principal lattice directions. For instance, in several common face-centered cubic (fcc) metals such as copper, nickel and platinum, the ground state structure of the self-interstitial is the split [100] interstitial structure, where two atoms are displaced in a positive and negative [100] direction from the lattice site. In body-centered cubic (bcc) iron the ground state interstitial structure is similarly a [110] split interstitial.

These split interstitials are often called dumbbell interstitials, because plotting the two atoms forming the interstitial with two large spheres and a thick line joining them makes the structure resemble a dumbbell weight-lifting device.

In other bcc metals than iron, the ground state structure is believed based on recent density-functional theory calculations to be the [111] crowdion interstitial, which can be understood as a long chain (typically some 10–20) of atoms along the [111] lattice direction, compressed compared to the perfect lattice such that the chain contains one extra atom.

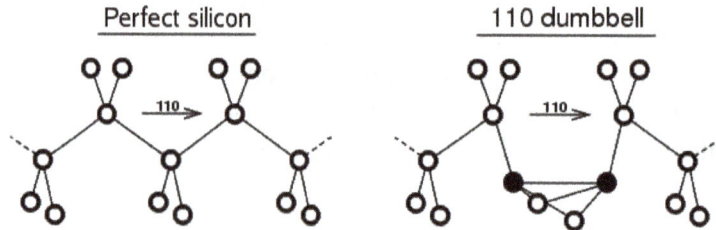

Structure of dumbbell self-interstitial in silicon. Note that the structure of
the interstitial in silicon may depend on charge state and doping level of the material.

In semiconductors the situation is more complex, since defects may be charged and different charge states may have different structures. For instance, in silicon, the interstitial may either have a split [110] structure or a tetrahedral truly interstitial one.

Carbon, notably in graphite and diamond, has a number of interesting self-interstitials - recently discovered using LDF-calculations is the "spiro-interestitial" in graphite, named after spiropentane, as the interstitial carbon atom is situated between two basal planes and bonded in a geometry similar to spiropentane.

Impurity Interstitials

Small impurity interstitial atoms are usually on true off-lattice sites between the lattice atoms. Such sites can be characterized by the symmetry of the interstitial atom position with respect to its nearest lattice atoms. For instance, an impurity atom I with 4 nearest lattice atom A neighbours (at equal distances) in an fcc lattice is in a tetrahedral symmetry position, and thus can be called a tetrahedral interstitial.

Large impurity interstitials can also be in split interstitial configurations together with a lattice atom, similar to those of the self-interstitial atom.

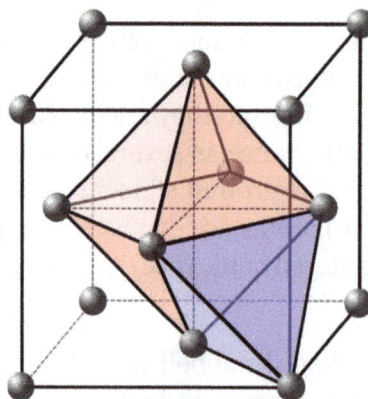

Octahedral (red) and tetrahedral (blue) interstitial symmetry polyhedra in a face-centered cubic lattice.
The actual interstitial atom would ideally be in the middle of one of the polyhedra.

Effects of Interstitials

Interstitials modify the physical and chemical properties of materials.

- Interstitial carbon atoms have a crucial role for the properties and processing of steels, in particular carbon steels.

- Impurity interstitials can be used e.g. for storage of hydrogen in metals.

- The amorphization of semiconductors such as silicon during ion irradiation is often explained by the buildup of a high concentration of interstitials leading eventually to the collapse of the lattice as it becomes unstable.

- Creation of large amounts of interstitials in a solid can lead to a significant energy buildup, which on release can even lead to severe accidents in certain old types of nuclear reactors (Wigner effect). The high-energy states can be released by annealing.

- At least in fcc lattice, interstitials have a large diaelastic softening effect on the material.

- It has been proposed that interstitials are related to the onset of melting and the glass transition.

Intrinsic Defects

An intrinsic defect is formed when an atom is missing from a position that must be filled in the crystal, creating a vacancy, or when an atom occupies an interstitial site where no atom would ordinarily appear, causing an interstitial. The two types of intrinsic point defects are shown in Figure.

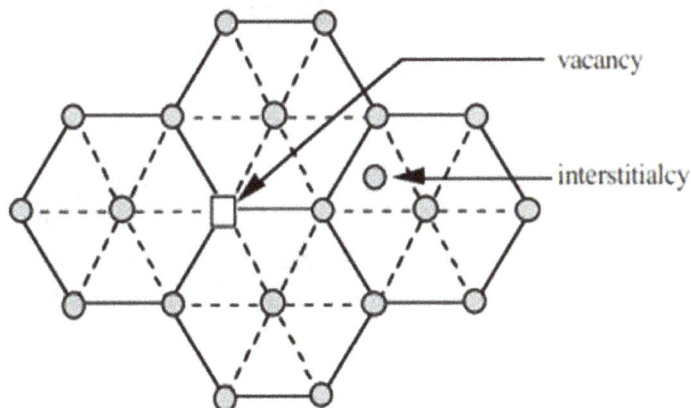

Illustration of a vacancy and an interstitial in a two-dimensional hexagonal lattice.

Because the interstitial sites in most crystalline solids are small (or have an unfavorable bonding configuration, as, for example, in the diamond lattice) interstitials are high-energy defects that are relatively uncommon. Vacancies, on the other hand, are present in a significant concentration in all crystalline materials. Their most pronounced effect is to govern the migration of atoms on the crystal lattice (solid state diffusion). In order for an atom to move easily from one crystal lattice site to another the target site must be vacant. As we shall see, the rate of diffusion on the crystal lattice is largely governed by the concentration of vacancies.

Ordered compounds can have more complex intrinsic defects. In most compounds the different species are charged to at least some degree. An intrinsic defect destroys the local charge balance, which must be restored in some way. The compound defects that preserve charge are easiest to visualize in binary ionic solids like NaCl. An isolated vacancy in an ionic solid creates an excess charge. The excess charge can be compensated by a paired vacancy on the sublattice of the other specie; for example, the excess charge associated with a Na vacancy is balanced if there is a Cl vacancy nearby. A neutral defect that involves paired vacancies on the cation and anion sub-lattices is called a Schottky defect. Alternatively, the charge imbalance caused by the vacancy can be corrected by adding an interstitial of the same specie; a Na vacancy is compensated by a Na interstitial. A neutral defect that is made up of a paired vacancy and interstitial is called a Frenkel defect. In compounds whose atoms are less strongly ionized it is energetically possible for species to exchange sites, so that an A-atom appears on the B sub-lattice or vice versa. This type of point defect is called an anti-site defect, and is fairly common in semiconducting compounds such as GaAs.

Extrinsic Defects

The extrinsic point defects are foreign atoms, which are called solutes if they are intentionally added to the material and are called impurities if they are not. The foreign atom may occupy a lattice sites, in which case it is called a substitutional solute (or impu- rity) or it may fill an interstitial site, in which case it is called an interstitial solute. Since the interstitial sites are relatively small, the type of the solute is largely determined by its size. Small atoms, such as hydrogen, carbon and nitrogen are often found in interstitial sites. Larger atoms are usually substitutional.

More complex extrinsic defects appear in compounds. If the valence of a substitutional defect in an ionic solid differs from that of the lattice ion then the excess charge is often compensated by a paired vacancy or interstitial. For example, when Mg^{++} ions are substituted for Na^+ in NaCl they tend to be paired with vacancies on the Na sublattice to maintain local charge neutrality. In semiconductors substitutional atoms with the wrong valence acts as electron donors or acceptors, as described below.

Extrinsic point defects affect almost all engineering properties, but they are particularly important in semiconducting crystals, where extrinsic defects are used to control electrical properties, and in structural metals and alloys, where extrinsic defects are added to increase mechanical strength. While these properties will be discussed later in the course, it is perhaps useful to identify the characteristics of the point defects that affect them.

Donors and Acceptors in Semiconductors

Point defects are intentionally added to semiconductors to control the type and concentration of charge carriers. Consider, for example, boron (valence 3) as a substi- tutional solute in elemental silicon. The saturated covalent bonds in silicon are shown schematically in Figure, and depend on the availability of four valence electrons per silicon atom. Since the bonds are saturated, silicon has very low conductivity in its pure state; pure silicon can only conduct electricity when electrons are excited into high- energy electron states. If boron is added, as in Figure, a valence electron is missing from the immediate environment of the boron atom, causing a hole in the bonding pattern. Electrons can then move from bond to bond by exchanging with the hole. The exchange requires some energy to separate the hole from the boron ion core, but this energy is small compared to that required to excite an electron from a Si-Si bond into a high-energy state. The room-temperature conductivity of Si increases significantly when a small amount of B is added. Electron-deficient solutes like boron that cause holes in the configuration of bonding electrons are called acceptors.

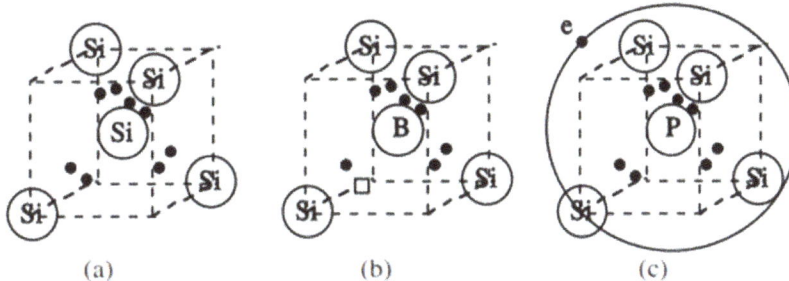

(a) Tetrahedral bonding configuration in Si. (b) Bonding around a B solute, showing a hole.
(c) Bonding around a P solute, showing an electron (e) in a loose excited orbital.

The conductivity also rises when a solute with an excess of electrons is added to a semiconductor with saturated bonds. For example, let phosphorous (valence 5) be added to Si, as in Figure. The 5 valence electrons of P are sufficient to fill the local covalent bonds with one electron left over. This electron can only go into an excited state, and or- bits about the P ion core somewhat as shown in the figure. It requires a relatively small energy increment to free this electron from the P core, in which case it can transport current by moving through the lattice. The conductivity of Si rises dramatically if a small amount of P is added. Electron-excess solutes such as P in Si are called donors.

Semiconductors whose electrical properties are controlled by electrically active solutes are called extrinsic semiconductors. Almost all of the semiconductors that are used in engineering devices are extrinsic.

Solution Hardening in Structural Materials

The addition of solute atoms almost always increases the mechanical strength of a solid. The phenomenon is called solution hardening. It is due to the fact that the solute atom is always a bit too large or a bit too small to fit perfectly into the crystal lattice site it is supposed to occupy, and distorts the crystal lattice in its attempt to fit as well as possible. As we shall see later, this distortion impedes the motion of the linear defects (dislocations) that are responsible for plastic deformation and, consequently, hardens the crystal. The distortion due to a substitutional solute is relatively small, though the associated hardening may be large enough to be useful in the engineering sense. The distortions due to interstitial atoms such as carbon and nitrogen are normally much greater because of the small size of the interstitial void in which they must fit. The hardening effect of interstitial solutes is large and technologically important; for example, high strength structural steels are alloys of Fe and C.

There is a simple crystallographic reason why interstitial solutes such as C are particularly effective in strengthening BCC metals such as Fe. The carbon atoms occupy octahedral interstitial sites in the BCC structure Since an atom in an octahedral void in BCC is closer to two of its neighbors that to the other four, it causes an asymmetric distortion of the lattice. As shown in Figure, the octahedron is stretched along its short axis, which is the a_3, or z-axis in the case shown in the figure. The asymmetric distortion of the interstitial site increases its interaction with the dislocations that cause plastic deformation and promotes hardening. In FCC alloys the interstitial sites are symmetric, and the lattice distortion is isotropic. Interstitial solutes are effective in hardening FCC alloys, but are less effective than in BCC alloys.

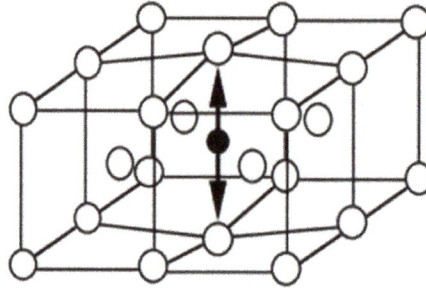

The local distortion of the BCC lattice by an interstitial atom in an O_z void.

Random Solid Solutions

At low concentrations it is usually reasonable to assume that the solutes are ran- domly distributed over the sites they occupy, in which case the material is called a random solid solution. The basic properties of a random substitutional solid solution can often be estimated by treating the material as if it were made up of pseudo-atoms that have the average properties of the components of the solution. In keeping with this idea, most of the properties of random substitutional solid solutions are simply proportional to the solute concentration. We shall see many examples in the following. A particular example is Vegard's Law, which asserts that when the solute concentration in a cubic solid solution is small its lattice parameter changes linearly with the concentration:

$$\frac{a - a_0}{a_0} = Kx$$

where a_0 is the lattice parameter of the crystal in pure form, x is the atom fraction of solute and K is a constant. The constant, K, is usually much larger for an interstitial solute because the solute atom must be fit into the small interstitial hole.

Tetragonal Distortion by Octahedral Interstitials in BCC

At larger solute concentrations the solute atoms interact and influence one another, and their ac- tual configuration in the crystal becomes important. An important and instructive example occurs in the case of octahedral interstitials in BCC materials, and is responsible for the technologically important fact that steel (Fe-C) that has a carbon content greater than a few tenths of an atom per- cent has a body-centered tetragonal rather than a BCC structure when it cooled rapidly (quenched) from elevated temperature. The body-centered tetragonal structure (BCT) differs from the BCC in that it is elongated along one of the three crystal axes (assume the a_3, or z-axis), as show in Figure.

The source of the BCT structure lies in the asymmetry of the octahedral void that is the preferred site for carbon. Since there are equal numbers of O_x, O_y and O_z octahedra in the BCC lattice, a ran- dom distribution of octahedral interstitials causes, on the average, an equal distortion of the three crystal axes and simply expands the BCC lattice. However, because of the asymmetric distortion, octahedral interstitials interact with one another and, if they are present in any significant concen- tration, preferentially occupy one of the three types of octahedral sites so that their short axes are aligned. When this happens there is a net long-range distortion of the crystal lattice in the direction of the short axis of the interstitial void that changes the structure from BCC to BCT.

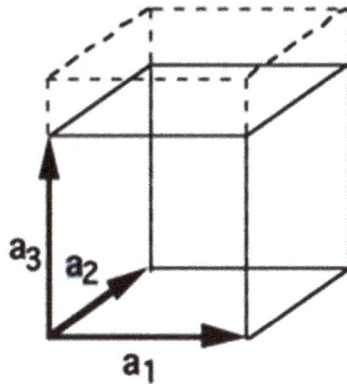

Tetragonal distortion of a cube. the distortion creates a tetragonal figure with an increased value of $|a_3|$.

When pure Fe is cooled from high temperature its structure changes from FCC to BCC. Carbon is more soluble in the FCC phase. If an FCC Fe-C alloy with a moderate carbon content is cooled very rapidly (quenched) then the carbon is trapped in the BCC product. The carbon interstitials preferentially adopt aligned octahedral sites, with the consequence that the quenched alloy, which is called martensite, has a body-centered tetragonal (BCT) structure. It is a very strong material with many important uses.

Decomposition and Ordering of Solid Solutions

Very few species are mutually soluble in all proportions. The solubility range is ordinarily limited by the preferential interaction of the solute atoms, and is limited in one of two ways. If the solute atoms bond preferentially to one another they tend to become associated in the solid solution, or cluster. As the concentration of solute increases the tendency to cluster becomes more pronounced, until the solution spontaneously decomposes into a mixture of two solutions, or phases, one of which is rich in the solute and one in the solvent. The point at which this decomposition occurs defines the solubility limit of the solid solution. On the other hand, if the solute atoms bond preferentially to the solvent then they tend to adopt ordered configurations in which solute and solvent alternate in a regular crystal pattern. When the solute concentration becomes great enough the solution decomposes into a mixture of a solution that is rich in solvent and an ordered compound that contains nearly stoichiometric proportions of the two atom types.

Note that the definition of a point defect in an ordered compound of species A and B is different from that in a solid solution of B in A. A point defect in a compound is a deviation from the perfectly ordered state of the compound structure. For example, in a solid solution of Au in Cu each Au atom constitutes a point defect, since the pure Cu crystal is the reference. In the intermetallic compound Cu_3Au, on the other hand, an Au atom is a point defect only if it occupies a Cu position in the Cu_3Au structure (in which case it is an anti-site defect), while a Cu atom is a point defect if it occupies an Au site. This re-emphasizes the important point that a "crystal defect" is a deviation from the configuration the crystal would have if it were perfectly ordered, and has no meaning until the reference state, the perfectly ordered crystal, is defined.

Schottky Defect

This is a defect which mainly arises if some of the lattice points are unoccupied. Such points which are unoccupied have been given the name lattice vacancies or 'holes'. The figure exhibits Schottky defect of crystals when existence of two holes, one due to a missing positive ion and the other due to missing negative ion in crystal lattice is there.

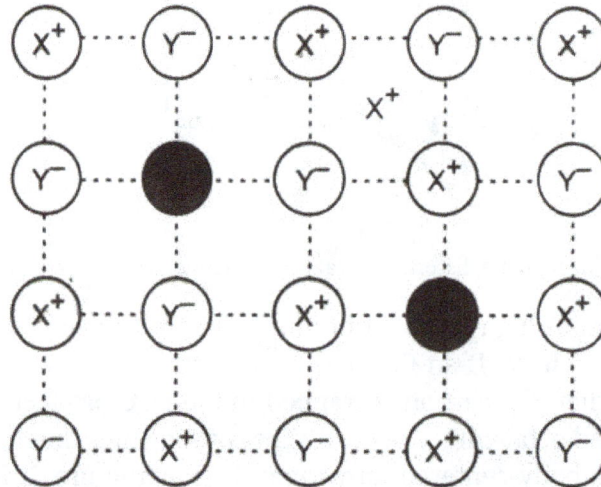

SCHOTTKY DEFECT OF CRYSTALS

It is also found that this defect is generally observed in strong ionic compounds having a high co-ordination number and the radius ratio r / R is not far below unity. Examples are cesium chloride and sodium chloride. Although both types of defects probably characterize crystals of non-stoichiometric compounds, the Schottky defects are more important.

Rees has introduced some symbolism for the constitution of imperfect crystals. He gave an idea that a lattice site appropriate presented. The nature of atom and type of site are specified for an occupied lattice; in this way an atom of type x on its proper lattice site is represented by x / o_x. The symbol x_{1-A} / o_x represents the fraction $1 - A(A < 1)$ the x lattice sites is occupied by correct species of atom. Now the fraction A remains vacant unless some other species of atom is specified as also located on x lattice sites.

Generally, interstitial positions are represented by Δ Hence interstitial positions occupied by a particular species of atom may then be symbolized by x / Δ With the help of this symbolism it is possible to know the concentration and nature of lattice defects in any system. The reactions by which lattice defects are formed can be represented by quasi chemical equations.

Lattice vacancies (holes) occur in almost all types of ionic solids. However Schottky defect appears more often than Frenkel defect. The reason is that the energy needed to form a Schottky defect is much less than that needed to form a Frankel defect.

Defect reaction in an oxide MO is written as

$$Null \ or \ 0 \rightleftharpoons V_O^{\bullet\bullet} + V_M''$$

Equilibrium constant for this reaction is

$$K_S = \left[V_O^{\bullet\bullet} \right]\left[V_M'' \right]$$

Here square brackets i.e. [ ] are used for concentration.

Equilibrium constant can be also be expressed as

$$K_S = \exp\left(-\frac{\Delta G_S}{RT} \right)$$

where ΔG_s is the molar free energy of defect formation and is ΔH_s - $T\Delta S_s$, where ΔH_s is the enthalpy for defect formation and ΔS_s is the entropy change which is mainly vibrational in nature and can be assumed to be constant. This leads to

$$K_S = \exp\left(-\frac{\Delta H_S}{RT} \right) \cdot \exp\left(-\frac{\Delta S_S}{RT} \right) = K_S \exp\left(-\frac{\Delta H_S}{RT} \right)$$

If Schottky defects dominate, then

$$\left[V_O^{\bullet\bullet} \right] = \left[V_M'' \right] = K_S^{1/2} = K_O^{1/2} \exp\left(-\frac{\Delta H_S}{2RT} \right)$$

Here, as one can see, defect concentrations are independent of pO_2.

Bound and Dilute Defects

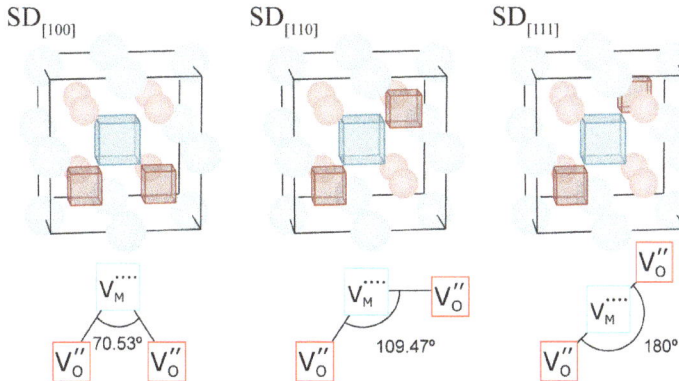

Three bound configurations of Schottky defects in an oxide with Fluorite structure.
Spheres represent atoms, cubes represent vacancies.

The vacancies that make up the Schottky defects have opposite charge, thus they experience a mutually attractive Coulomb force. If sufficient thermal energy is available the vacancies may migrate through the crystal lattice, and form bound clusters.

The bound clusters are typically less mobile than the dilute counterparts, as multiple species need to move in a concerted motion for the whole cluster to migrate. This has important implications for numerous functional ceramics used in a wide range of applications, including ion conductors, Solid oxide fuel cells and nuclear fuel.

Examples

This type of defect is typically observed in highly ionic compounds, highly coordinated compounds, and where there is only a small difference in sizes of cations and anions of which the compound lattice is composed. Typical salts where Schottky disorder is observed are NaCl, KCl, KBr, CsCl and AgBr. For engineering applications, Schottky defects are important in oxides with Fluorite structure, such as CeO_2, cubic ZrO_2, UO_2, ThO_2 and PuO_2.

Effect on Density

Since the total number of ions present in the crystal with this defect is less than the theoretical number of ions for a crystal of its volume, the density of the solid crystal is less than the theoretical density of the material.

Frenkel Defect

This defect generally arises when an ion occupies an interstitial position between lattice points.

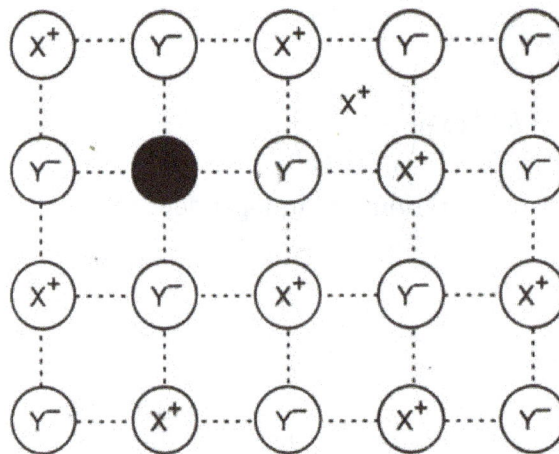

FRENKEL DEFECT OF CRYSTALS

Here positive ions occupy interstitial positions being smaller than negative ions. In the figure given above, it is clear that one of the positive ions occupies a position in interstitial space rather than at its own appropriate site in the lattice due to which a 'hole' is created in the lattice as shown in the figure.

The defect mostly appears in those compounds where positive and negative ions differ largely in their radii and coordination number is low.

The Frenkel Defect in a Molecule

The Frenkel Defect explains a defect in the molecule where an atom or ion (normally the cation) leaves its own lattice site vacant and instead occupies a normally vacant site. As depicted in the picture below, the cation leaves its own lattice site open and places itself between the area of all the other cations and anions. This defect is only possible if the cations are smaller in size when compared to the anions.

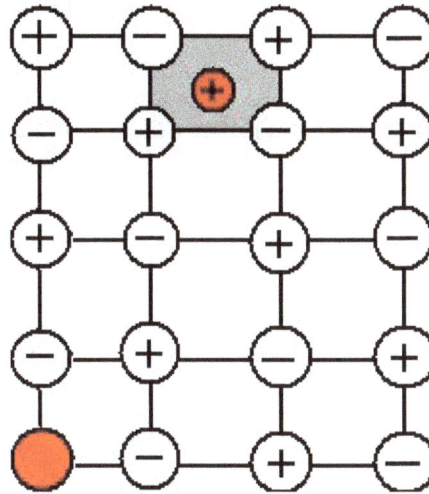

The Frenkel Defect in a molecule

The number of Frenkel Defects can be calculated using the equation:

$$\sqrt{NN^*}\, e^{\delta\frac{H}{2RT}}$$

where N is the number of normally occupied positions, N^* is the number of available positions for the moving ion, the delta H of formation is the enthalpy formation of one Frenkel defect, and R is the gas constant. Frenkel defects are intrinsic defects because the existence causes the Gibbs energy of a crystal to decrease, which means it's favorable to occur.

Molecules Found with a Frenkel Defect

The crystal lattices are relatively open and the coordination number is low.

For an oxide MO

$$M_M \rightleftharpoons M_i^{\bullet\bullet} + V_M{}''$$

which leads to

$$K_F = \frac{\left[M_i^{\bullet\bullet} \right]\left[V_M{}'' \right]}{\left[M_M \right]}$$

Or

$$K_F \left[M_M \right] = \left[M_i^{\bullet\bullet} \right]\left[V_M{}'' \right]$$

At reasonably low defect concentrations when= $[M_i^{\bullet\bullet}]$ and $[V_M{}''] \ll M_M$ and $M_M \approx 1$

Thus

$$[M_i^{\bullet\bullet}][V_M{}''] = K_F$$

i.e.

$$[M_i^{\cdot\cdot}]=[V_M^{''}]=K_F^{1/2}$$

In a similar manner what we did above for Schottky defects, one can now write

$$\left[M_i^{\bullet\bullet} \right]\left[V_M^{''} \right]=K_0^{1/2}\exp\left(-\frac{\Delta H_F}{2RT} \right)$$

Again we can see that the defect is independent of pO_2.

Examples

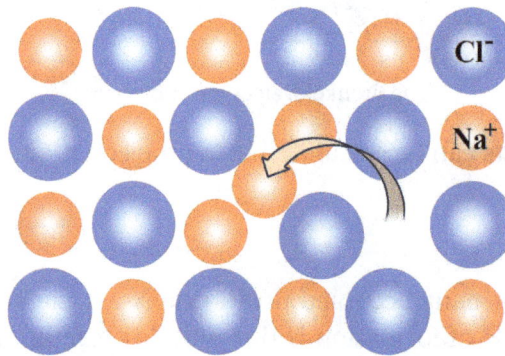

The Frenkel defect within the NaCl structure

Frenkel defects are exhibited in ionic solids with a large size difference between the anion and cation (with the cation usually smaller due to an increased effective nuclear charge)

Some examples of solids which exhibit Frenkel defects:

- Zinc sulfide,
- Silver(I) chloride,
- Silver(I) bromide (also shows Schottky defects),
- Silver(I) iodide.

These are due to the comparatively smaller size of Zn^{2+} and Ag^+ ions.

For example, consider a lattice formed by X^{n-} and M^{n+} ions. Suppose an M ion leaves the M sublattice, leaving the X sublattice unchanged. The number of interstitials formed will equal the number of vacancies formed.

One form of a Frenkel defect reaction in MgO with the oxide anion leaving the lattice and going into the interstitial site written in Kröger–Vink notation:

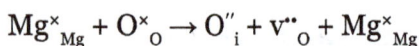

$$Mg^x_{Mg} + O^x_O \rightarrow O^{''}_i + v^{\cdot\cdot}_O + Mg^x_{Mg}$$

This can be illustrated with the example of the sodium chloride crystal structure. The diagrams below are schematic two-dimensional representations.

The defect-free NaCl structure

Two Frenkel defects within the NaCl structure

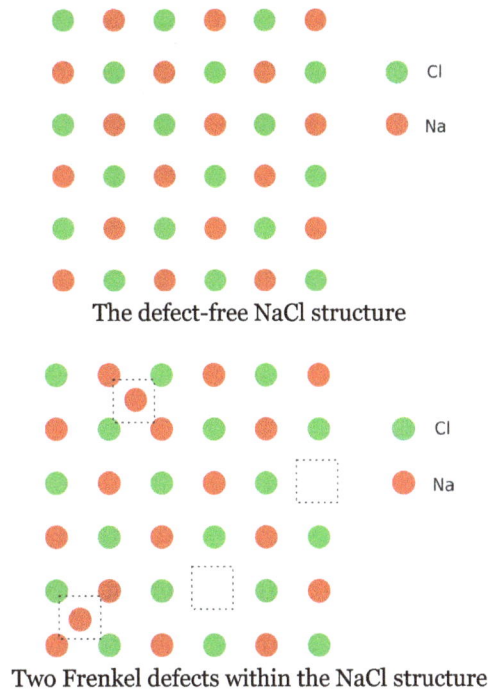

Schottky Defect vs Frenkel Defect

Schottky Defect

1. Equal number of cations and anions are missing from the lattice sites.

2. Found in highly ionic compounds with high coordination numbers and where the cations and anions are of similar size.

3. Density of the solid decreases.

Frenkel Defect

1. A cation leaves the normal lattice site and occupies an interstitial site.

2. Found in ionic compounds with low coordination numbers and where the anions are much larger in size than cations.

3. Density of the solid remains the same.

Coordination number: It is defined as the number of nearest neighbours of a particle in a closed packed structure.

Vacancy Defect

A vacancy is produced when an atom is missing from its original lattice site. So vacancy creates an empty lattice site as depicted below. Like other point defects, vacancy is also a zero-dimensional defect. Vacancy defect puts the neighboring atoms under tension. Due to the reduction in number of atoms in the crystalline solid, vacancy defect results in the reduction of density. However, hardness of the solid may increase.

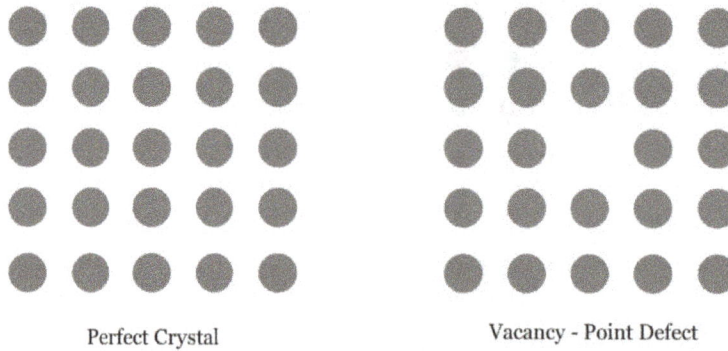

Perfect Crystal   Vacancy - Point Defect

This figure illustrates the difference between a perfect crystal and crystal structure with a vacancy defect.
Vacancy defect is caused by loss of one atom from its lattice site.

Where Vacancy Defects can be Found?

Vacancy defect can occur in any crystalline solid, in fact, it is inherent. Any material whose temperature is above absolute zero temperature (0K), can contain vacancies. With increase in temperature, number of vacancy defects increases exponentially.

Causes of Vacancy Defects in Solids:

Vacancy, a point defect, may occur due to various reasons, as enlisted below:

- For not allowing directional solidification, usually during casting.

- Due to increase in temperature of the solid for various processing like heat treatment, coating, etc. Number of vacancies in a specific amount of solid increases exponentially as indicated by the expression provided below.

- Due to irradiation or sputtering effect.

- Due to presence of residual tensile stress within the solid.

Effects of Vacancies in Solid:

- Although depends on material and its crystal structure, in general, vacancies can decrease the bulk modulus and can increase the Young's modulus.

- Large vacancy concentration can reduce the ductility of the crystalline solid; however, can increase the hardness.

- It can alter the thermal and electrical resistivity of the solid.

- Common physical properties, like melting point, color, etc. can also vary due to presence of vacancies.

Calculation of Number of Vacancies in Solid:

Vacancy defects are temperature sensitive. The number of vacancies present within a particular volume of solid at a particular temperature can be mathematically expressed by the following expression:

$$N_v = N \times e^{\left(\frac{-Q_v}{RT}\right)}$$

N_v = Total number of vacancies per unit volume of solid at a particular temperature (vacancies/m³).

N = Total number of lattice sites per unit volume of solid (lattice sites/m³).

Q_v = Energy required to produce single vacancy in that solid (J/mole).

R = Gas constant = 8.314 J/mole-K.

T = Absolute temperature of the solid (K).

Impurities Defect

Defects in ionic compounds because of replacement of ions by the ions of other compound is called impurities defects.

These defects arise when foreign atoms are present at the lattice site (in place of host atoms) or at the vacant interstitial sites. In the former case, we get substitutional solid solutions while in the latter case, we get interstitial solid solution. The formation of the former depends upon the electronic structure of the impurity while that of the later on the size of the impurity.

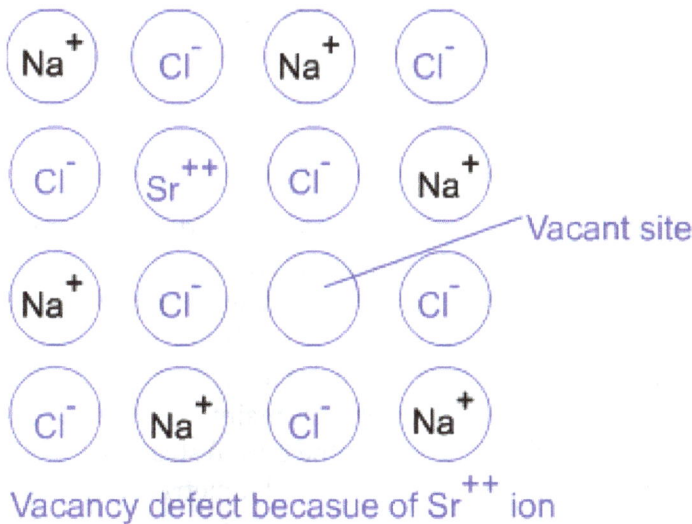

Vacancy defect becasue of Sr^{++} ion

In NaCl; during crystallization; a little amount of $SrCl_2$ is also crystallized. In this process, Sr^{++} ions get the place of Na+ ions and create impurities defects in the crystal of NaCl. In this defect, each of the Sr^{++} ion replaces two Na+ ions. Sr^{++} ion occupies one site of Na^+ ion; leaving other site vacant. Hence it creates cationic vacancies equal number of Sr^{++} ions. $CaCl_2$, AgCl, etc. also shows impurities defects.

These defects play an important role in semiconductors which are specially prepared for diodes, transistors, etc. Impurity atoms are present at the sites of regular parent atoms or in the interstitial spaces. These defects are two types.

Substitutional Impurities

This defect arises when an impurity atom replaces or substitutes parent atom in the crystal lattice. If the size of substitutional impurity is same as parent atom then the amount of strain around will be less, otherwise it will be more.

Ex: In extrinsic semiconductors either third or fifth group atoms occupy the sites of silicon or Germanium atoms.

Interstitial impurity: This defect arises when small sized foreign atom occupying an interstice or void space in the parent crystal without disturbing any of the parent atoms from their regular sites.

Ex: In steel, carbon atom (0.77r A°), being smaller in size, occupies interstitial position in Iron (2.250 r A°).

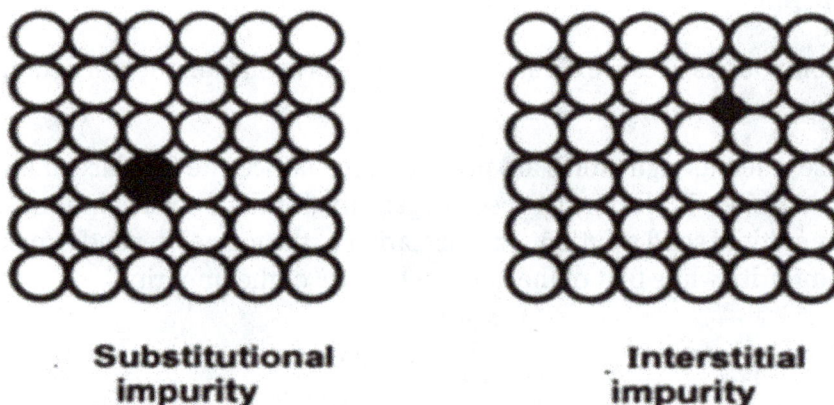

Substitutional
impurity

Interstitial
impurity

Non-stoichiometric Defects

There are many compounds in which the ratio of positive and negative ions present in the compound differs from that required by ideal formula of the compound. Such compounds are called non-stoichiometric or bertholide compounds. In these compounds composition may vary over a wide range. For example, in vanadium oxide (VO_x), x can be anywhere between 0.6 and 1.3. similarly FeO also exists as non-stoichiometric compound having composition $Fe_{0.92}$ to $Fe_{0.95}$O. In these compounds balance of positive and negative charges is maintained by having either extra electrons or extra positive charge. This results in defects in the crystal structure. The defects which disturb the stoichiometry of the compound are called non-stoichiometric defects. These defects are either due to the presence of excess metal or excess non-metal ions.

(a) Metal excess defects due to anion vacancies: a compound may have excess metal ion if a negative ion is absent from its lattice site, leaving a 'hole' which is occupied by electron to maintain electrical neutrally.

These types of defects are found in crystals which are likely to possess Schottky defects. Anion vacancies in alkali halides are produced by heating the alkali halides crystals in an atmosphere of alkali metal vapours. Under these conditions alkali metal atoms deposit on the surface and combine with metal ions. The electrons released during conversion of the metal atoms into ions diffuse into the crystals and occupy the sites vacated by anions.

The holes occupied by electrons are called F-centres (or color centres) and are responsible for the color of the compound and many other interesting properties. For example, the excess sodium in NaCl makes the crystal appear yellow, excess sodium in NaCl makes it violet and excess lithium ion LiCi makes it pink. Greater the number of F-centres, greater is the intensity of colour. Solids containing F-centres paramagnetic because the electrons occupying the 'holes' are unpaired.

(b) Metal excess defects due to interstitial cations: another way in which metal excess defect may occur is, if an extra positive ion is present in interstitial site. Electrical neutrality is maintained the presence of an electron in the interstitial site. This type of defects are exhibited in crystals which are likely to exhibit Frenkel defect, for example, when ZnO is heated, it loses oxygen reversibly. The excess metal is accommodated in interstitial with electrons tapped in the neighbourhood. The yellow and the electrical conductivity of the non stoichiometric ZnO are due to these trapped electrons.

(c) Metal deficiency due to cation vacancies: the non-stoichiometric compounds may have metal deficiency due to the absence of a metal ion in the lattice site. The charge is balanced by an adjacent ion having higher positive charge. These types of defects are generally shown by compounds of transit metals. For example, non-stoichiometric cuprous oxide (Cu_2O) can be prepared in laboratory. In this oxide copper to oxygen ratio is slightly less than 2:1. This is due to the reason that some of the positions which were to be occupied by Cu^+ ion are vacant whereas some positions are occupied by Cu^{2+} ions. Another example of this type is non-stoichiometric FeO which is mostly with a composition $Fe_{0.95}O$. in the crystals of FeO some Fe^{2+} ions are missing and the loss of positive charged balanced by the presence of required number of ions.

Line Defects

Dislocations are linear defects; they are lines through the crystal along which crystallographic registry is lost.

Dislocations are another type of defect in crystals. Dislocations are areas were the atoms are out of position in the crystal structure. Dislocations are generated and move when a stress is applied. The motion of dislocations allows slip – plastic deformation to occur.

Before the discovery of the dislocation by Taylor, Orowan and Polyani in 1934, no one could figure out how the plastic deformation properties of a metal could be greatly changed by solely by forming (without changing the chemical composition). This became even bigger mystery when in the early 1900's scientists estimated that metals undergo plastic deformation at forces much smaller than the theoretical strength of the forces that are holding the metal atoms together. Many metallurgists remained skeptical of the dislocation theory until the development of the transmission electron microscope in the late 1950's. The TEM allowed experimental evidence to be collected that showed that the strength and ductility of metals are controlled by dislocations.

There are two basic types of dislocations, the edge dislocation and the screw dislocation. Actually, edge and screw dislocations are just extreme forms of the possible dislocation structures that can occur. Most dislocations are probably a hybrid of the edge and screw forms but this discussion will be limited to these two types.

Edge Dislocations

The edge defect can be easily visualized as an extra half-plane of atoms in a lattice. The dislocation is called a line defect because the locus of defective points produced in the lattice by the dislocation lie along a line. This line runs along the top of the extra half-plane.

The inter-atomic bonds are significantly distorted only in the immediate vicinity of the dislocation line.

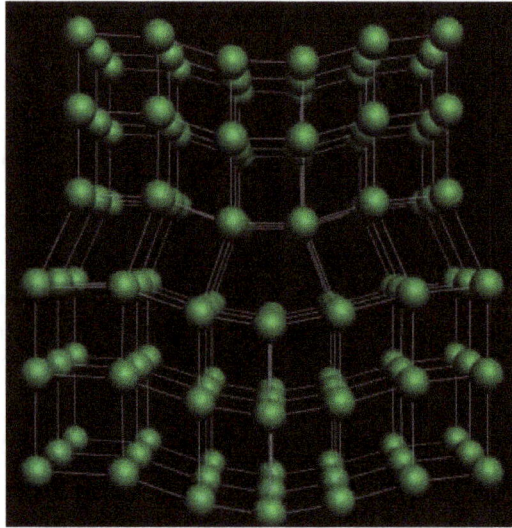

Understanding the movement of a dislocation is key to understanding why dislocations allow deformation to occur at much lower stress than in a perfect crystal. Dislocation motion is analogous to movement of a caterpillar. The caterpillar would have to exert a large force to move its entire body at once. Instead it moves the rear portion of its body forward a small amount and creates a hump. The hump then moves forward and eventual moves all of the body forward by a small amount.

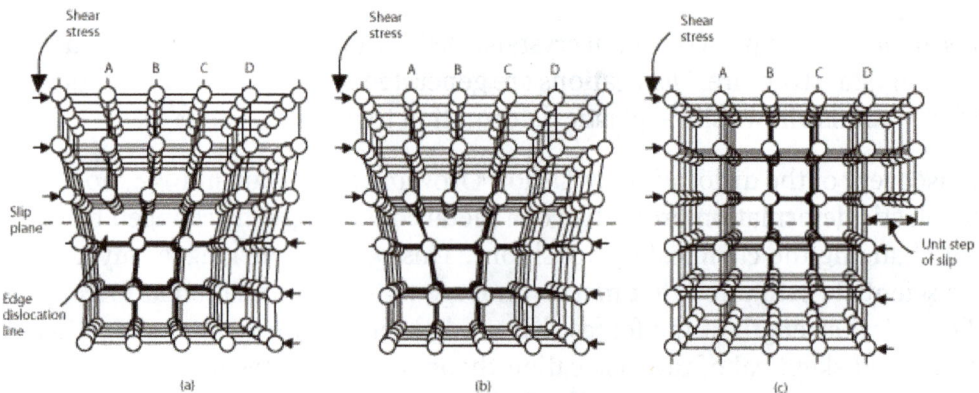

As shown in the set of images above, the dislocation moves similarly moves a small amount at a time. The dislocation in the top half of the crystal is slipping one plane at a time as it moves to the right from its position in image (a) to its position in image (b) and finally image (c). In the process of slipping one plane at a time the dislocation propagates across the crystal. The movement of the dislocation across the plane eventually causes the top half of the crystal to move with respect to the bottom half. However, only a small fraction of the bonds are broken at any given time. Movement in this manner requires a much smaller force than breaking all the bonds across the middle plane simultaneously.

Screw Dislocations

There is a second basic type of dislocation, called screw dislocation. The screw dislocation is slightly more difficult to visualize. The motion of a screw dislocation is also a result of shear stress, but the defect line movement is perpendicular to direction of the stress and the atom displacement, rather than parallel. To visualize a screw dislocation, imagine a block of metal with a shear stress applied across one end so that the metal begins to rip. This is shown in the upper right image. The lower right image shows the plane of atoms just above the rip. The atoms represented by the blue circles have not yet moved from their original position. The atoms represented by the red circles have moved to their new position in the lattice and have reestablished metallic bonds. The atoms represented by the green circles are in the process of moving. It can be seen that only a portion of the bonds are broke at any given time. As was the case with the edge dislocation, movement in this manner requires a much smaller force than breaking all the bonds across the middle plane simultaneously.

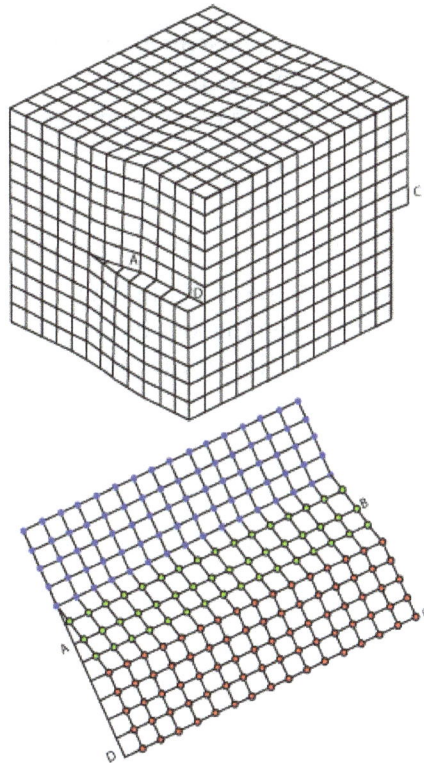

If the shear force is increased, the atoms will continue to slip to the right. A row of the green atoms will find there way back into a proper spot in the lattice (and become red) and a row of the blue atoms will slip out of position (and become green). In this way, the screw dislocation will move upward in the image, which is perpendicular to direction of the stress. Recall that the edge dislocation moves parallel to the direction of stress. As shown in the image below, the net plastic deformation of both edge and screw dislocations is the same, however.

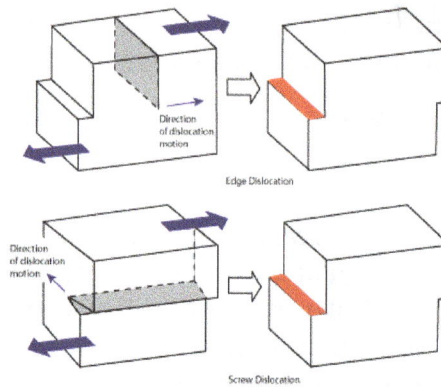

Edge Dislocation

Screw Dislocation

The dislocations move along the densest planes of atoms in a material, because the stress needed to move the dislocation increases with the spacing between the planes. FCC and BCC metals have many dense planes, so dislocations move relatively easy and these materials have high ductility. Metals are strengthened by making it more difficult for dislocations to move. This may involve the introduction of obstacles, such as interstitial atoms or grain boundaries, to "pin" the dislocations. Also, as a material plastically deforms, more dislocations are produced and they will get into each others way and impede movement. This is why strain or work hardening occurs.

In ionically bonded materials, the ion must move past an area with a repulsive charge in order to get to the next location of the same charge. Therefore, slip is difficult and the materials are brittle. Likewise, the low density packing of covalent materials makes them generally more brittle than metals.

Jogs and Kinks Dislocation

Jogs and kinks are atomic scale changes in direction of a dislocation line. Jogs are out of the slip plane and kinks lie in the dislocations slip plane. For the edge dislocation shown, the jogs are of opposite sign (direction) and the region between the jogs has moved up from the original slip plane by one atomic spacing. This may be caused by the adsorption of vacancies on the extra half plane.

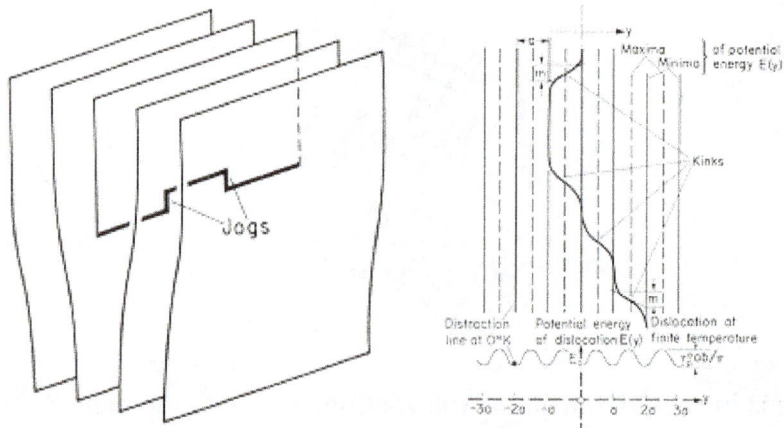

A slip plane has low energy locations for the dislocation line separated by barriers of higher energy. At 0 K the dislocation will tend to lie in just one minimum energy location. At finite temperature, the dislocation line can move via thermal fluctuations and may occupy several distinct minima along its length. The line section crossing the barriers between these minima is the kink. For both jogs and kinks the Burgers vector and line vector are those of the dislocation, the line vector being along the direction of the jog or the kink. These features may therefore be of a different type (edge, screw) than the dislocation and may restrict dislocation glide.

Planar Defects

Planar defect is an imperfection in form of a plane between uniform parts of the material. The most important planar defect is a grain boundary. Formation of a boundary between two grains may be imagined as a result of rotation of crystal lattice of one of them about a specific axis. Depending on the rotation axis direction, two ideal types of a grain boundary are possible:

- Tilt boundary – rotation axis is parallel to the boundary plane;

- Twist boundary - rotation axis is perpendicular to the boundary plane:

- An actual boundary is a "mixture" of these two ideal types.

Grain boundaries are called large-angle boundaries if misorientation of two neighboring grains exceeds 10°-15°.

Grain boundaries are called small-angle boundaries if misorientation of two neighboring grains is 5° or less.

Grains, divided by small-angle boundaries are also called subgrains.

Grain boundaries accumulate crystal lattice defects (vacancies, dislocations) and other imperfections, therefore they effect on the metallurgical processes, occurring in alloys and their properties.

Since the mechanism of metal deformation is a motion of crystal dislocations through the lattice, grain boundaries, enriched with dislocations, play an important role in the deformation process.

Diffusion along grain boundaries is much faster, than throughout the grains.

Segregation of impurities in form of precipitating phases in the boundary regions causes a form of corrosion, associated with chemical attack of grain boundaries. This corrosion is called Intergranular corrosion.

Stacking Faults and Twin Boundaries

A disruption of the long-range stacking sequence can produce two other common types of crystal defects: 1) a stacking fault and 2) a twin region. A change in the stacking sequence over a few atomic spacings produces a stacking fault whereas a change over many atomic spacings produces a twin region.

A stacking fault is a one or two layer interruption in the stacking sequence of atom planes. Stacking faults occur in a number of crystal structures, but it is easiest to see how they occur in close packed structures. For example, it is know from a previous discussion that face centered cubic (fcc) structures differ from hexagonal close packed (hcp) structures only in their stacking order. For hcp and fcc structures, the first two layers arrange themselves identically, and are said to have an AB arrangement. If the third layer is placed so that its atoms are directly above those of the first (A) layer, the stacking will be ABA. This is the hcp structure, and it continues ABABABAB. However it is possible for the third layer atoms to arrange themselves so that they are in line with the first layer to produce an ABC arrangement which is that of the fcc structure. So, if the hcp structure is going along as ABABAB and suddenly switches to ABABABCABAB, there is a stacking fault present.

Alternately, in the fcc arrangement the pattern is ABCABCABC. A stacking fault in an fcc structure would appear as one of the C planes missing. In other words the pattern would become ABCAB-CAB_ABCABC.

If a stacking fault does not corrects itself immediately but continues over some number of atomic spacings, it will produce a second stacking fault that is the twin of the first one. For example if the stacking pattern is ABABABAB but switches to ABCABCABC for a period of time before switching back to ABABABAB, a pair of twin stacking faults is produced. The red region in the stacking sequence that goes ABCABCACBACBABCABC is the twin plane and the twin boundaries are the A planes on each end of the highlighted region.

Grain Boundaries in Polycrystals

Another type of planer defect is the grain boundary. Up to this point, the discussion has focused on defects of single crystals. However, solids generally consist of a number of crystallites or grains. Grains can range in size from nanometers to millimeters across and their orientations are usually rotated with respect to neighboring grains. Where one grain stops and another begins is know as a grain boundary. Grain boundaries limit the lengths and motions of dislocations. Therefore, having smaller grains (more grain boundary surface area) strengthens a material. The size of the grains can be controlled by the cooling rate when the material cast or heat treated. Generally, rapid cooling produces smaller grains whereas slow cooling result in larger grains.

Grain Boundary

The juncture between adjacent grains is called a grain boundary. The grain boundary is a transition region in which some atoms are not exactly aligned with either grain.

Therefore, the grain boundaries are:

- Where grains meet in a solid.

- Transition regions between the neighboring crystals.

- Where there is a disturbance in the atomic packing.

Characterization of grain boundaries is of critical importance in materials studies. The properties of grain boundaries often determine the grains':

- Formation

- Evolution

- Stabilization (or dissolution)

Grain boundaries have two types, as per their orientation:

- Low-angle grain boundaries are those with a misorientation less than about 11 degrees.

- High-angle grain boundaries are whose misorientation is greater than about 11 degrees.

High-angle boundaries are considerably more disordered, with large areas of poor fit and a comparatively open structure. The mobility of low-angle boundaries is much lower than that of high-angle boundaries. Both low- and high-angle boundaries are retarded by grain refinement, to minimize or prevent recrystallization or grain growth during heat treatment.

The grain-boundary atoms are more easily and rapidly dissolved, or corroded, than the atoms within the grains. Grain boundaries will oxidize or corrode more rapidly, usually referred to as grain-boundary penetration or intergranular attack. In under-deposit corrosion and hydrogen damage, grain boundaries are the site at which the methane collects that leads to the intergranular cracking characteristic of hydrogen damage.

High and Low Angle Boundaries

It is convenient to categorize grain boundaries according to the extent of misorientation between the two grains. *Low-angle grain boundaries* (LAGBs) or *subgrain boundaries* are those with a misorientation less than about 15 degrees. Generally speaking they are composed of an array of dislocations and their properties and structure are a function of the misorientation. In contrast the properties of *high-angle grain boundaries* (HAGBs), whose misorientation is greater than about 15 degrees (the transition angle varies from 10–15 degrees depending on the material), are normally found to be independent of the misorientation. However, there are 'special boundaries' at particular orientations whose interfacial energies are markedly lower than those of general HAGBs.

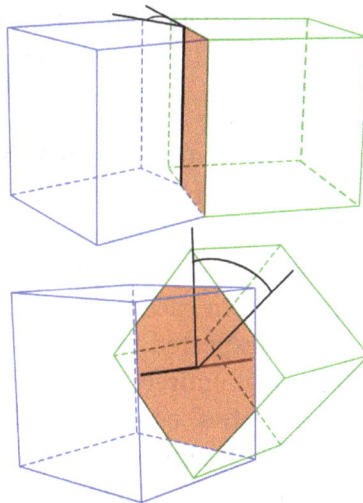

Schematic representations of a tilt boundary (top) and a twist boundary between two idealised grains.

The simplest boundary is that of a tilt boundary where the rotation axis is parallel to the boundary plane. This boundary can be conceived as forming from a single, contiguous crystallite or grain which is gradually bent by some external force. The energy associated with the elastic bending of the lattice can be reduced by inserting a dislocation, which is essentially a half-plane of atoms that act like a wedge, that creates a permanent misorientation between the two sides. As the grain is bent further, more and more dislocations must be introduced to accommodate the deformation resulting in a growing wall of dislocations – a low-angle boundary. The grain can now be considered to have split into two sub-grains of related crystallography but notably different orientations.

An alternative is a twist boundary where the misorientation occurs around an axis that is perpendicular to the boundary plane. This type of boundary incorporates two sets of screw dislocations. If the Burgers vectors of the dislocations are orthogonal, then the dislocations do not strongly interact and form a square network. In other cases, the dislocations may interact to form a more complex hexagonal structure.

These concepts of tilt and twist boundaries represent somewhat idealized cases. The majority of boundaries are of a mixed type, containing dislocations of different types and Burgers vectors, in order to create the best fit between the neighboring grains.

If the dislocations in the boundary remain isolated and distinct, the boundary can be considered to be low-angle. If deformation continues, the density of dislocations will increase and so reduce the spacing between neighboring dislocations. Eventually, the cores of the dislocations will begin to overlap and the ordered nature of the boundary will begin to break down. At this point the boundary can be considered to be high-angle and the original grain to have separated into two entirely separate grains.

In comparison to LAGBs, high-angle boundaries are considerably more disordered, with large areas of poor fit and a comparatively open structure. Indeed, they were originally thought to be some form of amorphous or even liquid layer between the grains. However, this model could not explain the observed strength of grain boundaries and, after the invention of electron microscopy, direct evidence of the grain structure meant the hypothesis had to be discarded. It is now accepted that a boundary consists of structural units which depend on both the misorientation of the two grains and the plane of the interface. The types of structural unit that exist can be related to the concept of the *coincidence site lattice*, in which repeated units are formed from points where the two misoriented lattices happen to coincide.

In coincident site lattice (CSL) theory, the degree of fit (Σ) between the structures of the two grains is described by the reciprocal of the ratio of coincidence sites to the total number of sites. In this framework, it is possible to draw the lattice for the 2 grains and count the number of atoms that are shared (coincidence sites), and the total number of atoms on the boundary (total number of site). For example, when $\Sigma=3$ there will be one atom each 3 that will be shared between the two lattices. Thus a boundary with high Σ might be expected to have a higher energy than one with low Σ. Low-angle boundaries, where the distortion is entirely accommodated by dislocations, are $\Sigma1$. Some other low-Σ boundaries have special properties, especially when the boundary plane is one that contains a high density of coincident sites. Examples include coherent twin boundaries (e.g., $\Sigma3$) and high-mobility boundaries in FCC materials (e.g., $\Sigma7$). Deviations from the ideal CSL orientation may be accommodated by local atomic relaxation or the inclusion of dislocations at the boundary.

Describing a Boundary

A boundary can be described by the orientation of the boundary to the two grains and the 3-D rotation required to bring the grains into coincidence. Thus a boundary has 5 macroscopic degrees of freedom. However, it is common to describe a boundary only as the orientation relationship of the neighbouring grains. Generally, the convenience of ignoring the boundary plane orientation, which is very difficult to determine, outweighs the reduced information. The relative orientation of the two grains is described using the rotation matrix:

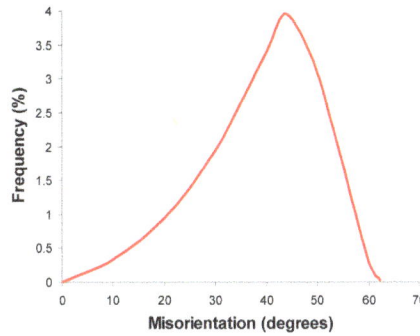

The characteristic distribution of boundary misorientations in a completely randomly oriented set of grains for cubic symmetry materials.

$$R = \begin{bmatrix} a_{11} & a_{12} & a_{13} \\ a_{21} & a_{22} & a_{23} \\ a_{31} & a_{32} & a_{33} \end{bmatrix}$$

Using this system the rotation angle θ is:

$$2\cos\theta + 1 = a_{11} + a_{22} + a_{33}$$

while the direction [uvw] of the rotation axis is:

$$[(a_{32} - a_{23}), (a_{13} - a_{31}), (a_{21} - a_{12})]$$

The nature of the crystallography involved limits the misorientation of the boundary. A completely random polycrystal, with no texture, thus has a characteristic distribution of boundary misorientations. However, such cases are rare and most materials will deviate from this ideal to a greater or lesser degree.

Boundary Energy

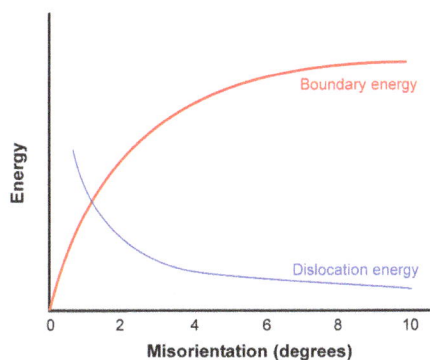

The energy of a tilt boundary and the energy per dislocation as the misorientation of the boundary increases.

The energy of a low-angle boundary is dependent on the degree of misorientation between the neighbouring grains up to the transition to high-angle status. In the case of simple *tilt boundaries* the energy of a boundary made up of dislocations with Burgers vector b and spacing h is predicted by the Read-Shockley equation:

$$\gamma_s = \gamma_0 \theta (A - \ln \theta)$$

where:

$$\theta = b / h$$

$$\gamma_0 = Gb / 4\pi (1 - \nu)$$

$$A = 1 + \ln(b / 2\pi r_0)$$

with G is the shear modulus, ν is Poisson's ratio, and r_0 is the radius of the dislocation core. It can be seen that as the energy of the boundary increases the energy per dislocation decreases. Thus there is a driving force to produce fewer, more misoriented boundaries (i.e., grain growth).

The situation in high-angle boundaries is more complex. Although theory predicts that the energy will be a minimum for ideal CSL configurations, with deviations requiring dislocations and other energetic features, empirical measurements suggest the relationship is more complicated. Some predicted troughs in energy are found as expected while others missing or substantially reduced. Surveys of the available experimental data have indicated that simple relationships such as low Σ are misleading:

It is concluded that no general and useful criterion for low energy can be enshrined in a simple geometric framework. Any understanding of the variations of interfacial energy must take account of the atomic structure and the details of the bonding at the interface.

Excess Volume

The excess volume is another important property in the characterization of grain boundaries. Excess volume was first proposed by Bishop in a private communication to Aaron and Bolling in 1972. It describes how much expansion is induced by the presence of a GB and is thought that the degree and susceptibility of segregation is directly proportional to this. Despite the name the excess volume is actually a change in length, this is because of the 2D nature of GBs the length of interest is the expansion normal to the GB plane. The excess volume (δV) is defined in the following way,

$$\delta V = \left(\frac{\partial V}{\partial A} \right)_{T, p, n_i} ,$$

at constant temperature T, pressure p and number of atoms n_i. Although a rough linear relationship between GB energy and excess volume exists the orientations where this relationship is violated can behave significantly differently affecting mechanical and electrical properties.

Experimental techniques have been developed which directly probe the excess volume and have been used to explore the properties of nanocrystalline copper and nickel. Theoretical methods have also been developed and are in good agreement. A key observation is that there is an inverse relationship with the bulk modulus meaning that the larger the bulk modulus (the ability to compress a material) the smaller the excess volume will be, there is also direct relationship with the lattice constant, this provides methodology to find materials with a desirable excess volume for a specific application.

Boundary Migration

The movement of grain boundaries (HAGB) has implications for recrystallization and grain growth while subgrain boundary (LAGB) movement strongly influences recovery and the nucleation of recrystallization.

A boundary moves due to a pressure acting on it. It is generally assumed that the velocity is directly proportional to the pressure with the constant of proportionality being the mobility of the boundary. The mobility is strongly temperature dependent and often follows an Arrhenius type relationship:

$$M = M_0 \exp\left(-\frac{Q}{RT}\right)$$

The apparent activation energy (Q) may be related to the thermally activated atomistic processes that occur during boundary movement. However, there are several proposed mechanisms where the mobility will depend on the driving pressure and the assumed proportionality may break down.

It is generally accepted that the mobility of low-angle boundaries is much lower than that of high-angle boundaries. The following observations appear to hold true over a range of conditions:

- The mobility of low-angle boundaries is proportional to the pressure acting on it.

- The rate controlling process is that of bulk diffusion.

- The boundary mobility increases with misorientation.

Since low-angle boundaries are composed of arrays of dislocations and their movement may be related to dislocation theory. The most likely mechanism, given the experimental data, is that of dislocation climb, rate limited by the diffusion of solute in the bulk.

The movement of high-angle boundaries occurs by the transfer of atoms between the neighbouring grains. The ease with which this can occur will depend on the structure of the boundary, itself dependent on the crystallography of the grains involved, impurity atoms and the temperature. It is possible that some form of diffusionless mechanism (akin to diffusionless phase transformations such as martensite) may operate in certain conditions. Some defects in the boundary, such as steps and ledges, may also offer alternative mechanisms for atomic transfer.

Since a high-angle boundary is imperfectly packed compared to the normal lattice it has some amount of *free space* or *free volume* where solute atoms may possess a lower energy. As a result, a boundary may be associated with a *solute atmosphere* that will retard its movement. Only at higher velocities will the boundary be able to break free of its atmosphere and resume normal motion.

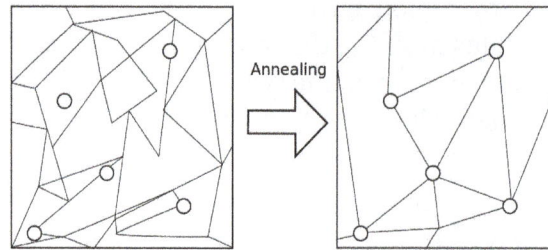

Grain growth can be inhibited by second phase particles via Zener pinning.

Both low- and high-angle boundaries are retarded by the presence of particles via the so-called Zener pinning effect. This effect is often exploited in commercial alloys to minimise or prevent recrystallization or grain growth during heat-treatment.

Complexion

Grain boundaries are the preferential site for segregation of impurities, which may form a thin layer with a different composition from the bulk. For example, a thin layer of silica, which also contains impurity cations, is often present in silicon nitride. These grain boundary phases are thermodynamically stable and can be considered as quasi-two-dimensional phase, which may undergo to transition, similar to those of bulk phases. In this case structure and chemistry abrupt changes are possible at a critical value of a thermodynamic parameter like temperature or pressure. This may strongly affect the macroscopic properties of the material, for example the electrical resistance or creep rates. Grain boundaries can be analyzed using equilibrium thermodynamics but cannot be considered as phases, because they do not satisfy Gibbs'definition: they are inhomogeneous, may have a gradient of structure, composition or properties. For this reasons they are defined as complexion: an interfacial material or stata that is in thermodynamic equilibrium with its abutting phases, with a finite and stable thickness (that is typically 2–20 Å). A complexion need the abutting phase to exist and its composition and structure need to be different from the abutting phase. Contrary to bulk phases, complexions also depend on the abutting phase. For example, silica rich amorphous layer present in Si_3N_3, is about 10 Å thick, but for special boundaries this equilibrium thickness is zero. Complexion can be grouped in 6 categories, according to their thickness: monolayer, bilayer, trilayer, nanolayer (with equilibrium thickness between 1 and 2 nm) and wetting. In the first cases the thickness of the layer will be constant; if extra material is present it will segregate at multiple grain junction, while in the last case there is no equilibrium thickness and this is determined by the amount of secondary phase present in the material. One example of grain boundary complexion transition is the passage from dry boundary to biltilayer in Au-doped Si, which is produced by the increase of Au.

Effect to the Electronic Structure

Grain boundaries can cause failure mechanically by embrittlement through solute segregation but they also can detrimentally affect the electronic properties. In metal oxides it has been shown theoretically that at the grain boundaries in Al_2O_3 and MgO the insulating properties can be significantly diminished. Using density functional theory computer simulations of grain boundaries have shown that the band gap can be reduced by up to 45%. In the case of metals grain boundaries increase the resistivity as the size of the grains relative to the mean free path of other scatters becomes significant.

Stacking Fault

All crystals whose structures can be described by layers are prone to stacking faults. A stacking fault is any defect that alters the periodic sequence of layers. These defects may be a wrong layer inserted into the sequence, a change of the layer sequence or a different translation between two subsequent layers. These defects may affect the whole crystal or a finite region if e.g. an additional layer is present between an otherwise perfect sequence of layers. DISCUS contains a tool to create layered crystal structures and to introduce stacking faults into these crystals. The crystals are formed in a two step procedure. First, the origin and type of each layer is determined and second, the atoms corresponding to each layer are introduced into the crystal. The user can define each layer type, the translation vectors between consecutive layers and the correlation between neighbouring layers. A new feature of the stacking fault part of DISCUS is the addition of rotational disorder for the layer sequences.

The stacking fault sequence is defined by several parameters that can be set in the 'stack' sublevel of DISCUS:

- *Type of layers :*

  The positions of all atoms within a layer are read from a *DISCUS* type structure file. These layer files have to be created for each layer type involved beforehand using the various *DISCUS* tools.

- *Translations :*

  A translation vector between neighbouring layers of each type must be provided. Thus for N different layer types result in a N*N matrix of translation vectors. An example for translation vectors in a cubic face centered structure is given in table.

- *Uncertainties for translation vectors:*

  In some materials small deviations in the translation vectors might occur. This behaviour can be simulated in *DISCUS* by setting a standard deviation σ to each of the elements of the translation vector matrix. *DISCUS* will calculate the actual translation vector as sum of the 'ideal' vector plus a Gaussian distributed part defined by the value of σ.

- *Correlations :*

  A correlation matrix is used to define the probabilities of two layer types to be nearest neighbours. No further correlations are taken into account.

- *Crystal shape :*

  The resulting crystal can be generated using two different modes: First, the crystal continuously grows in one direction as given by the translation vector(s). Secondly one or two coordinates can be constrained to a finite range, which results in a zig-zag shaped crystal. If any of the parameters is not equal to zero, the corresponding coordinate of the origin is taken modulo this parameter. Note, that *DISCUS* does not check whether the moduli vectors are translation vectors of the current space group.

Table: Translation vectors for stacking faults in a cubic face centered structure

Layer type A	Layer type B	Translation vector
A	A	$(1, 1, 1)$
A	B	$\frac{1}{2}(1, 1, 0)$
A	C	$\frac{1}{2}(1, 0, 1)$
B	A	$\frac{1}{2}(\bar{1}, \bar{1}, 0)$
B	B	$(1, 1, 1)$
B	C	$\frac{1}{2}(0, 1, 1)$
C	A	$\frac{1}{2}(\bar{1}, 0, \bar{1})$
C	B	$\frac{1}{2}(0, \bar{1}, \bar{1})$
C	C	$(1, 1, 1)$

The command 'create' in the 'stack' segment of *DISCUS* creates the list of layer origins and 'run' actually generates the corresponding crystal by decorating the origins with the individual layer types. In order to speed up the calculation of the Fourier transform, rather than using the resulting complete structure, the scattering intensity is calculated in the following way. The scattering density $\rho(\mathbf{r})$ of a layered structure can be expressed as the scattering density of the individual layer types convoluted with the layer origin distribution.

$$\rho(\mathbf{r}) = \sum_{i=1}^{nl} \left\{ \sum_{j=1}^{no} o_{ij}(\mathbf{r}) \right\} \cdot l_i(\mathbf{r})$$

The outer sum runs over all nl layer types and the inner sum runs over all origins o_{ij} of layer type i. The variable li is the scattering density of layer type i. Using the convolution theorem, the Fourier transform of this expression becomes

$$F\{\rho(\mathbf{r})\} = \sum_{i=1}^{nl} F\left\{ \sum_{j=1}^{no} o_{ij}(\mathbf{r}) \right\} \cdot F\{l_i(r)\}$$

Here F denotes the Fourier transform. This procedure not only speeds up the calculation but it also allows the usage of much larger crystal sizes since the actual structure does not have to be created in order to calculate the Fourier transform.

A simple example of the stacking fault segment of *DISCUS* will be given here. The resulting layered structure is shown in figure. The filled circles represent atoms of layer type A whereas the empty circles stand for atoms of layer type B. The layers are stacked in y-direction and are shifted in x-direction by ±0.5 lattice units between the layers. The structure consists of preferred AB and BA sequences, however AA and BB pairs are present as well.

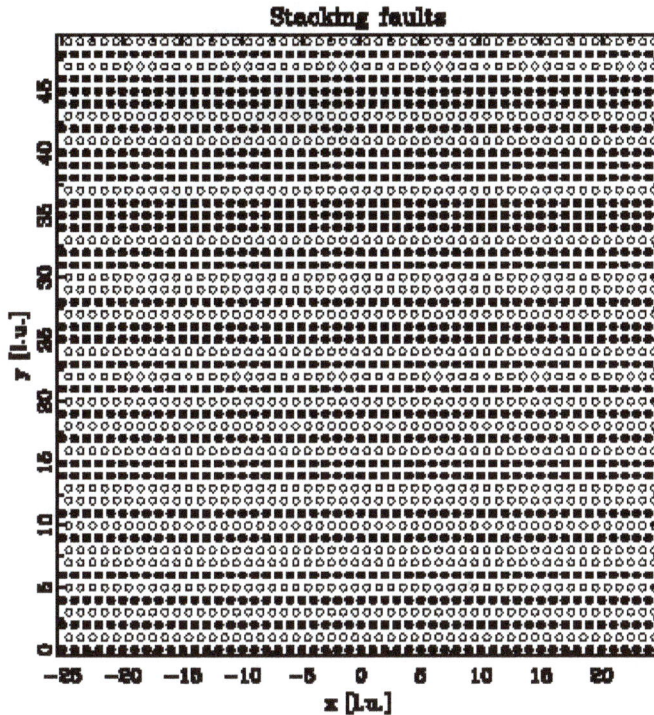

Example of stacking faults

The following *DISCUS* macro file was used to generate the layers structure shown in figure.

Stacking Faults in Semiconductors

Many compound semiconductors, e.g. those combining elements from groups III and V or from groups II and VI of the periodic table, crystallize in the fcc zincblende or hcp wurtzite crystal structures. In a semiconductor crystal, the fcc and hcp phases of a given material will usually have different band gap energies. As a consequence, when the crystal phase of a stacking fault has a lower band gap than the surrounding phase, it forms a quantum well, which in photoluminescence experiments leads to light emission at lower energies (longer wavelengths) than for the bulk crystal. In the opposite case (higher band gap in the stacking fault), it constitutes an energy barrier in the band structure of the crystal that can affect the current transport in semiconductor devices.

The Concept of a Generalized Stacking Fault

(or γ-surface) was introduced in the 1960's by Vitek on bcc metals. The misfit energy across a shear plane is derived by displacing the two half of a crystal of a given quantity in that plane and then by allowing atoms to relax normal to the shear plane. The process is then repeated so as to cover any shear direction in a given plane. For each energy obtained during a displacement, we remove the energy of the crystal without displacement. The energy difference is finally divided by the surface of the plane. Then the entire energy surface is generated. The example is 2D (displacements along one direction only in a given plane) for easiest visualisation, the result is a gamma-line along [010] in (100).

Supercell and successive steps for gamma-line calculation.
A rigid body shear is applied (at 30 GPa here) on Mg-Pv along [010] and in (100).

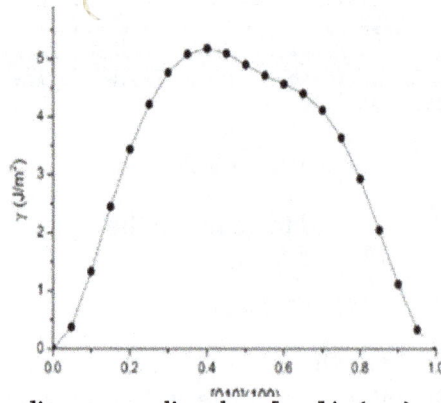

Corresponding gamma- line along [010] in (100), at 30 GPa.

Examples: γ-surfaces in MgSiO$_3$ perovskite

are calculated by atomistic potential calculations. The latters are performed using Buckingham pairwise potentials as implemented in the LAMMPS code. We present here the gamma-surfaces for (100), (010), (001), (110), (101) and (011) obtained at 30 GPa.

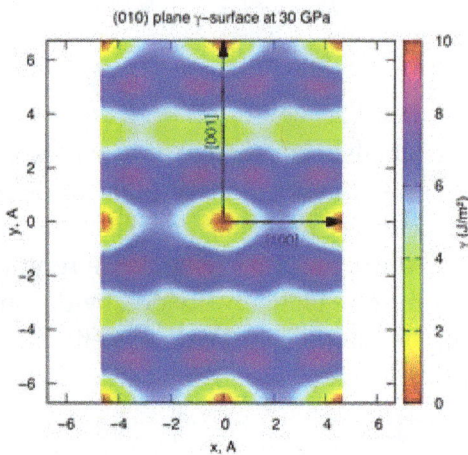

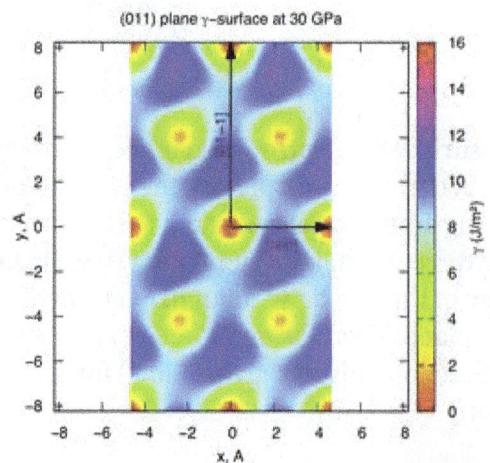

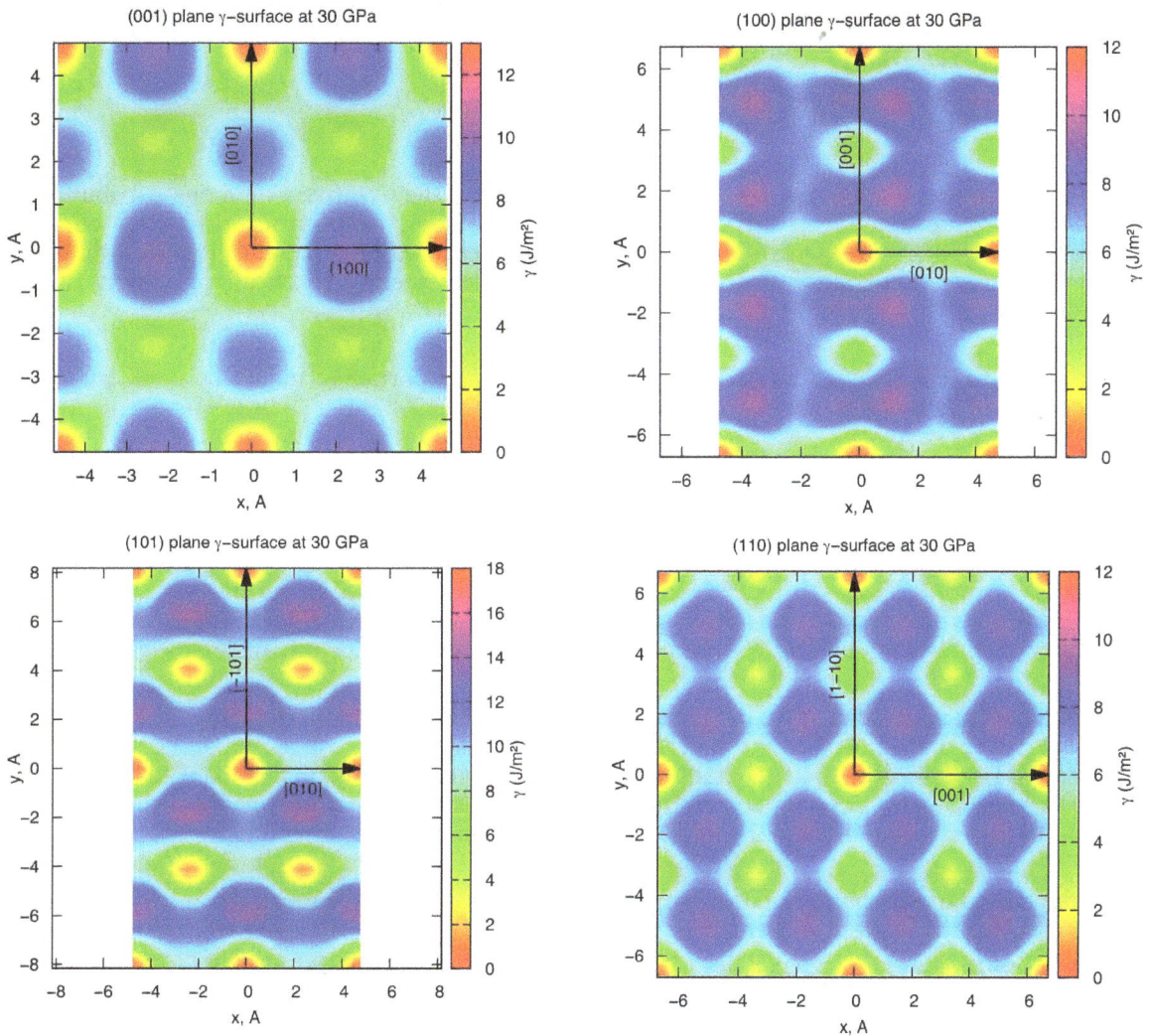

(001) plane γ–surface at 30 GPa

(100) plane γ–surface at 30 GPa

(101) plane γ–surface at 30 GPa

(110) plane γ–surface at 30 GPa

References

- Bean, Jonathan J.; McKenna, Keith P. (2016). "Origin of differences in the excess volume of copper and nickel grain boundaries". Acta Materialia. 110: 246–257. doi:10.1016/j.actamat.2016.02.040

- Chiang, Yet-Ming; Birnie III, Dunbar; Kingery, W. David (1997). Physical Ceramics: Principles for Ceramic Science and Engineering (1st ed.). John Wiley & Sons. pp. 102–107. ISBN 0-471-59873-9

- Frenkel, Yakov (1926). "Über die Wärmebewegung in festen und flüssigen Körpern (About the thermal motion in solids and liquids)". Zeitschrift für Physik. Springer. 35 (8): 652–669. Bibcode:1926ZPhy...35..652F. doi:10.1007/BF01379812. Retrieved 21 October 2015

- Aaron, H. B.; Bolling, G. F. (1972). "Free volume as a criterion for grain boundary models". Surface Science. 31 (C): 27–49. Bibcode:1972SurSc..31...27A. doi:10.1016/0039-6028(72)90252-X

- Schilling, W. (1978). "Self-interstitial atoms in metals". Journal of Nuclear Materials. 69–70: 465. Bibcode:1978JNuM...69..465S. doi:10.1016/0022-3115(78)90261-1

Permissions

All chapters in this book are published with permission under the Creative Commons Attribution Share Alike License or equivalent. Every chapter published in this book has been scrutinized by our experts. Their significance has been extensively debated. The topics covered herein carry significant information for a comprehensive understanding. They may even be implemented as practical applications or may be referred to as a beginning point for further studies.

We would like to thank the editorial team for lending their expertise to make the book truly unique. They have played a crucial role in the development of this book. Without their invaluable contributions this book wouldn't have been possible. They have made vital efforts to compile up to date information on the varied aspects of this subject to make this book a valuable addition to the collection of many professionals and students.

This book was conceptualized with the vision of imparting up-to-date and integrated information in this field. To ensure the same, a matchless editorial board was set up. Every individual on the board went through rigorous rounds of assessment to prove their worth. After which they invested a large part of their time researching and compiling the most relevant data for our readers.

The editorial board has been involved in producing this book since its inception. They have spent rigorous hours researching and exploring the diverse topics which have resulted in the successful publishing of this book. They have passed on their knowledge of decades through this book. To expedite this challenging task, the publisher supported the team at every step. A small team of assistant editors was also appointed to further simplify the editing procedure and attain best results for the readers.

Apart from the editorial board, the designing team has also invested a significant amount of their time in understanding the subject and creating the most relevant covers. They scrutinized every image to scout for the most suitable representation of the subject and create an appropriate cover for the book.

The publishing team has been an ardent support to the editorial, designing and production team. Their endless efforts to recruit the best for this project, has resulted in the accomplishment of this book. They are a veteran in the field of academics and their pool of knowledge is as vast as their experience in printing. Their expertise and guidance has proved useful at every step. Their uncompromising quality standards have made this book an exceptional effort. Their encouragement from time to time has been an inspiration for everyone.

The publisher and the editorial board hope that this book will prove to be a valuable piece of knowledge for students, practitioners and scholars across the globe.

Index

www.ingramcontent.com/pod-product-compliance
Lightning Source LLC
Chambersburg PA
CBHW080245230326
41458CB00097B/3340